ECONOMIC CONTROL STRUCTURES
A non-Walrasian Approach

CONTRIBUTIONS TO ECONOMIC ANALYSIS

188

Honorary Editor:
J. TINBERGEN

Editors:
D. W. JORGENSON
J. WAELBROECK

NORTH-HOLLAND
AMSTERDAM • NEW YORK • OXFORD • TOKYO

ECONOMIC CONTROL STRUCTURES
A non-Walrasian Approach

Béla MARTOS
Institute of Economics
Hungarian Academy of Sciences
Budapest, Hungary

1990

NORTH-HOLLAND
AMSTERDAM • NEW YORK • OXFORD • TOKYO

ELSEVIER SCIENCE PUBLISHERS B.V.
Sara Burgerhartstraat 25
P.O. Box 211, 1000 AE Amsterdam, The Netherlands

Distributors for the United States and Canada:

ELSEVIER SCIENCE PUBLISHING COMPANY INC.
655 Avenue of the Americas
New York, N.Y. 10010, U.S.A.

Typeset using the facilities of
the University of Illinois at Chicago
Office of Publications Services

ISBN: 0 444 87411 9

© ELSEVIER SCIENCE PUBLISHERS B.V., 1990

All rights reserved. No part of this publication may be reproduced, stored in a retrieval system, or transmitted, in any form or by any means, electronic, mechanical, photocopying, recording or otherwise, without the prior written permission of the publisher, Elsevier Science Publishers B.V./ Physical Sciences and Engineering Division, P.O. Box 1991, 1000 BZ Amsterdam, The Netherlands.

Special regulations for readers in the U.S.A. - This publication has been registered with the Copyright Clearance Center Inc. (CCC), Salem, Massachusetts. Information can be obtained from the CCC about conditions under which photocopies of parts of this publication may be made in the U.S.A. All other copyright questions, including photocopying outside of the U.S.A., should be referred to the publisher.

No responsibility is assumed by the publisher for any injury and/or damage to persons or property as a matter of products liability, negligence or otherwise, or from any use or operation of any methods, products, instructions or ideas contained in the material herein.

PRINTED IN THE NETHERLANDS

Introduction to the Series

This series consists of a number of hitherto unpublished studies, which are introduced by the editors in the belief that they represent fresh contributions to economic science.

The term "economic analysis" as used in the title of the series has been adopted because it covers both the activities of the theoretical economist and the research worker.

Although the analytical methods used by the various contributors are not the same, they are nevertheless conditioned by the common origin of their studies, namely theoretical problems encountered in practical research. Since for this reason, business cycle research and national accounting, research work on behalf of economic policy, and problems of planning are the main sources of the subjects dealt with, they necessarily determine the manner of approach adopted by the authors. Their methods tend to be "practical" in the sense of not being too far remote from application to actual economic conditions. In addition they are quantitative rather than qualitative.

It is the hope of the editors that the publication of these studies will help to stimulate the exchange of scientific information and to reinforce international cooperation in the field of economics.

The Editors

PREFACE

This monograph grew out of research work I did over more than a decade with the purpose of adopting a new approach to the theory of economic mechanisms. Economies which are not in and do not tend to Walrasian equilibrium were the focus of this study.

Although I have published several papers on the subject — some of which are available only in Hungarian — this book is not a newly edited collection of them. The basic line of my thinking has remained the same, but the models used to support my argument have been critically revised. More general results have been achieved by the use of more appropriate mathematical tools, and the addition of new model structures completed the discussion. Nonetheless, it is to be admitted that my results, even if taken together with those of my fellows-in-arms who worked on adjacent fields, have made possible, at most, a modest progress in the direction of a more general theory of economic mechanisms.

I have explained the economic message with the help of both formal mathematics and verbal argumentation which run in parallel. The reader who feels that his understanding is not helped by the mathematical formalism may restrict his attention to Chapters 1 to 4, 14, 18 and 19, from which I tried to eliminate formulae. From these chapters an incomplete but general insight may be gained.

The inspiration for this research work came from *János Kornai*. His effect on my thinking can be discerned not only in our joint papers and the jointly edited volume *Non-Price Control* but in almost everything I have written since the publication of his pioneering work *Anti-Equilibrium*. I cannot reciprocate, with however many references and finely stylized acknowledgements, the help he offered me in the form of advice, criticism and thought provoking discussion.

Many discerning comments on this work were provided by *Mária Augusztinovics*, *Antal Mátyás* and *András Simonovits*, as well as other readers and critics *András Bródy*, *István Dancs*, *Zsuzsa Kapitány*, *János Kovács*, *Anna Lee*, and *Márton Tardos*. Allow me to acknowledge here their expert helpfulness. Mrs. *Teréz Zimányi* did a fine job in deciphering and typing my almost illegible handwriting.

I have taken the risk of writing this monograph in English being constantly aware that my flat style cannot measure up to the standards of a demanding reader. I am very much indebted to *Simon J. Corrigan* who took on the thankless job of correcting my English. If the book is now readable the credit must go to him, but I am to blame for any awkwardness which remains.

B. M.

CONTENTS

Preface vii

Notation xiii

PART ONE: INTRODUCTION TO ECONOMIC SYSTEMS AND MATHEMATICAL CONTROL THEORY

1. Motivations, goals, instruments 3

2. The economic system 11

3. The economic control: principles and experiences 23

4. Control structures: vegetative functioning and coordination 31

5. Systems control 43

6. State equations of the transfer element and the control circuit 53

7. Stability and viability 63

Bibliographical notes to Part One 79

PART TWO: THE CONTROL OF AN OPEN LEONTIEF-ECONOMY WITHOUT COORDINATION

8. The common real sphere and the common form of the behavioural equations 83

9. The stock signal (Model S) 97

10. The order signal (Model B) 113

11. The combination of stock and order signals (Model SB) 129

12. The commercial stock signal (Model C) 137

13. The supply-side price signal (Model P) 147

14. Summary of Part Two 159

Bibliographical notes to Part Two 167

PART THREE: EQUIVALENT CONTROLLERS, PARTIAL COORDINATION

15. The Laplace transformation 173

16. Equivalence of controllers 181

17. Partially coordinated equivalent controllers (Model E) 191

18. Summary of Part Three 207

19. After-word: equilibria with rationing reconsidered 215

Bibliographical notes to Part Three 231

REFERENCES 233

INDEX 237

LIST OF FIGURES

1. Dissection of a controller 33
2. Open-loop control 45
3. Feedback (closed-loop) control 47
4. Viability concepts 76
5. Block diagram of Model S 99
6. Block diagram of Model B 118
7. Locus of the destabilizing eigenvalues of **A** in Model B 122
8. Block diagram of Model SB 131
9. Block diagram of Model C 141
10. Block diagram of Model P 152
11. Block diagram of rationing 225

NOTATION

1 *The use of different types*

Real or complex scalar quantities (numbers) are denoted by small case Greek letters, column vectors by small case Latin letters and matrices by Latin capitals: β, b, B. Greek capitals are used with different signification in different places. Small case Latin letters used as subscripts or superscripts denote non-negative integers. Sets are denoted by script-type capitals: $\mathcal{F}$.

Constant vectors and matrices depending neither on time t nor on the Laplace-variable s are set in block letters (different from those used in the text): x, X. Time-dependent vectors and matrices are set in italics: x, X; their Laplace-transforms in bold face type: **x**, **X**. In this way the argument t or s can often be omitted.

2 *Symbols*

$\dot{x} := dx/dt$: the dot above a variable is the sign of the derivative with respect to time

x_0 : the subscript $_0$ denotes the initial value (at time t_0) of a variable, $x_0 := x(t_0)$

x_i : subscripts distinguish different values of a variable

x^i, X^{ij} : superscripts refer to components of vectors or entries of matrices

x^* : the asterisk denotes the normal or reference value of a variable

x', X' : the prime is the sign of transposition, x' is the transpose (row vector) of column vector x

$<x>$: the diagonal matrix formed from the vector x

:= : equals by definition: "is defined as"

=: : equals by definition: "is denoted by"

$\overset{\circ}{X}$: the interior of the set X

$|\alpha|$: the absolute value (modulus) of the real or complex number α

$|B|$: the determinant of the square matrix B

$\|b\|$, $\|B\|$: the norm of vector b or square matrix B

col B : the column vector formed from matrix B by writing its columns one underneath the other

3 Standard use of certain letters

Generally a letter as a mathematical notation keeps its meaning within a chapter, but between different chapters the meaning may change. The letters in the following list, however, keep their meaning throughout the book.

A : the matrix of material input coefficients on current account

c : the end use vector

$C := (E - A)^{-1}$, the Leontief-inverse of A

e_i : the i-th unit vector $[0, 0, \ldots, 1^{(i)}, \ldots, 0]'$

e : the summation vector $[1, 1, \ldots, 1]'$

E : the unit matrix $[e_1, e_2, \ldots, e_k]$

$\mathcal{F}$: the spectrum (set of eigenvalues) of matrix A

g : value added (profitability indicator)

K : stock of backlog orders

n : the number of products, sectors, producers

n_0 : the number of zeroes in A

p : the price vector

q : the vector of output (product) stocks

r : the production vector

$\bar{r}$: productive capacity

V : the matrix of input (material) stocks

W : the matrix of placement of orders

Y : the matrix of commodity transfers

β : scalar parameter (damping exponent)

γ : scalar parameter (natural frequency)

$\Gamma := 2\beta\gamma \cdot d(\cdot)/dt + \gamma^2 \cdot (\cdot)$, a differential operator

$\varepsilon = 2.71828\ldots$

$\iota = \sqrt{-1}$

$\varphi \in \mathcal{F}$: an eigenvalue of **A**

ρ : the spectral radius of **A**

4 The system of cross-references

The book consists of 19 chapters arranged in three parts. The subtitles, definitions, theorems, remarks and formulae are consecutively numbered within each chapter without making distinction between them. These reference numbers are set in boldface type. Thus the number **19** refers to item **19** of the current chapter, 3.**19** to item **19** in Chapter 3.

References to books and papers are made by the last name of the author(s) followed by the year of publication, e.g. ARROW - HAHN (1971), with the two following exceptions: AE for *Anti-Equilibrium*, KORNAI (1971); NPC for *Non-Price Control*, KORNAI - MARTOS (1981). In order to avoid excessive or repeated references within the main text, most are relegated to special sections called *Bibliographical Notes* at the end of the Parts.

PART ONE

INTRODUCTION TO ECONOMIC SYSTEMS AND MATHEMATICAL CONTROL THEORY

In this introductory part of the treatise I will deal with two sorts of issues. In the first four chapters I will outline certain basic ideas from theoretical economics which serve as a background to my studies. In the three subsequent chapters I will recall a series of basic concepts and theorems from the mathematical theory of automatic control in a form suited to the analysis carried out in the rest of the book. The connecting of these two doctrines will lead to the substantial results to be found in the second and third parts of the book.

Chapter 1

MOTIVATIONS, GOALS, INSTRUMENTS

1 *A foretaste*

The housewife goes to the grocer's in order to buy sugar (and other edibles) for her family. This elementary economic process has been described in various ways by different schools of economic theory:

a) The housewife takes into account the price of sugar (and of other goods) and the amount of money she can spend. Thereafter she selects a bundle among the possible variants, namely that which is most favourable to her, and buys as much sugar as prescribed by this optimum solution. If there is not enough sugar in store, then the price of the sugar is raised (let us not dwell now on who does it) such that the growing supply and falling demand become equal. (General Equilibrium Theory.)

b) The above process modifies so that if there is not enough sugar, then the housewife gets less (perhaps none at all), the sugar is "rationed", but the price will not be changed. The housewife either buys other food which is not rationed or else her money remains unspent. (The Theory of Equilibria with Rationing)

c) The housewife takes stock of sugar in her pantry and observes the speed with which it diminishes. On this basis she decides how much sugar to buy now, without being influenced by today's sugar price. If she fails to get this amount she places an order (or waits), buys something else, searches about, grumbles. (The Theory of Quantity Adjustment.)

In contrast with description a) which is called "Walrasian", descriptions b) and c) can be called "non-Walrasian" and we will soon see that the latter better deserves this adjective than the former.

In the same spirit as our description of the behaviour of the household as buyer in three different conceptions, we could apply the same process as regards the household as seller (of manpower, say), to the productive firm both as buyer and seller of goods and services, the tradesman and so on, with the difference that on the seller's side it is the unrealizable intention to sell (unemployment, excess capacity, stockpiling of products) which marks the non-Walrasian state of the economy.

I will return repeatedly to the comparison of these theories (also to features which have not yet been mentioned); here I only wanted to give a foretaste. In my treatise I apply the third approach mentioned above and do this in the following conviction:

a) There are processes in real-life economy such that this approach describes them appropriately.

b) In the course of the past development of economic theory the quantity adjustment has not been given the weight and attention it deserves, and the formal description.of such controllers is conspicuously absent.

c) There are a number of phenomena both in capitalistic and socialistic economies which can be better analysed within this framework than within its predecessors.

In the wording of the above sentences I have been deliberately cautious, remaining aware of the fact that this theory is just beginning to take shape and has not yet gone further than trying to explain a small fraction of economic processes. Even the introductory example of the sugar-buying housewife is misleading on two counts. Firstly I chose an everyday, routine household decision where one could easily suggest that, in the case of slowly changing consumer prices, the buyer would not pay much attention to them. It actually transpired in the course of a conversation with three housewives, that they had all bought bananas a short time before (bananas are not consistently available in Hungary) but none of them actually knew the price. Still the example would have been less convincing in the case of a decision as to the purchase of a house or even a washing machine.

Secondly, in what follows I will pay much more attention to the interrelations between firms — this being the very subject of my work — than to the household. And insensitivity to prices may well be more characteristic of firms than of households. In the case of state-owned or state-protected firms, sensitivity to prices may be weakened to such an extent that quantity adjustment becomes predominant (KORNAI, 1988).

I hope I have made clear that I am not offering any "general theory". But while stressing that in the control of economic processes price signals and monetary controls play an indispensable role, it might still be useful to examine which kinds of processes can be controlled without these signals and which cannot. Consequently, I insist that it is not possible to construct a theory of economic mechanisms without taking prices and monetary processes into account. But I also insist that the neglect of quantity signals, which has become an unfortunate tradition in economic theory, is no less an omission.

In my imagination a "general theory" emerges as a future synthesis of these different approaches. But this has not yet been attempted.

2 Is there a choice of economic mechanisms?

This monograph deals with the problem of how the operation of the individual economic agents — and thereby of the economy as a whole — is, or can be, controlled. The first part of this question, "how the economy *is* controlled", is as old as economic theory itself. But the second half of the question, how an economy *can be* controlled, is meaningful only under the assumption that the society has a choice between different mechanisms.

For orthodox economists (whether orthodox liberals or orthodox communists) the question does not arise. For them the question of economic policy within a given mechanism is an actual problem of choice. The rules of the game, the functioning of the given system, are also worth studying in order to apply this knowledge to the improvement of economic policy-making. But the basic structure of the dominating control is never questioned by them. It remains as it has developed in the course of the historical progress of capitalism, or, at the other end of the spectrum, as was established by the Soviets under the leadership of Lenin and Stalin. But for reformers of East and West alike the question of choice among economic mechanisms within a given ownership system remains open, and therefore structures which can be potentially useful are worth studying.

Thus it is no wonder that the research I am doing has been motivated by the debate on and experimentation with economic mechanisms, which already has been underway for three decades in Hungary (and with less vehemence and continuity in other socialist countries).

Even so, what I am writing is not a contribution to this debate. No analysis, proposition or recommendation can be found in this book concerning the reform of the control of a socialist economy. (As a matter of fact, the abstract economy I am dealing with is one with highly developed division of labour but the ownership of the means of production is not specified.) Nonetheless, it is my ambition (as well that of colleagues who preceded or worked alongside me) to explore elements, building blocks and structural forms which may be relevant to the study of actual economic mechanisms.

This kind of research builds from elements of real life (*from that which exists*). Only a small number of these elements are taken into account simultaneously; many are suppressed or treated with oversimplification. However, among the selected elements and phenomena not only those interrelations are studied whose existence is empirically validated, but also such as *bear potentiality*, to the effect that neither common sense nor experience excludes the existence of a given structure, property or relation. (A theoretical physicist is applauded for "discovering" on paper a particle whose existence is only confirmed experimentally at a later date. This would be a curiosity in economics. The physicist must be more confident in the explanatory power of his theory — and rightly so.) In a favourable case such a train of thoughts may show what is worth testing and what is not. It is evident that this kind of research has neither the intention nor the power to make proposals: *what to do*.

As a result of the above reasoning — which I do not wish the reader to accept *a priori*, but hope to convince him of by the results which follow — I set out from the following working hypotheses:

a) Society does have a choice between elements and structures from which to build an economic mechanism and between the ways they are integrated into a viable entity.

b) The theoretical foundations of this choice can be supported by research which studies these potential structures in an abstract mathematical setting.

c) Even within such a framework, it is required that the structures in question consist of elements existing in real life and relationships which can be (or better *have been*) observed, but full empirical verification can be neither required nor obtained, if only because of numerous omissions.

3 Control theory, economic policy and control structures

The mathematical concepts and tools of the theory of automatic control will be applied in my analysis. Application of control theory (especially the theory of optimal control) to economics is a fashionable research trend. But this monograph lies far from the mainstream of that school as regards both the questions asked and the instruments applied in answering them.

Namely in the optimum control framework the problem is the following: the existing economic mechanism being *given*, to form a mathematical model thereof, and with its help to calculate the values (or trajectories) of the control parameters which are in some sense optimal, and are such that by using these values as the instruments of economic policy the operation of the economy can be regulated. In this way the task is to determine the optimal economic policy (or optimal plan) and to elaborate the models and procedures for these calculations. If you leaf through a bibliography on the economic applications of control theory (e.g. DERAKHSHAN, 1978), conference proceedings (e.g. JANSSEN - PAU - STRASZAK, 1981) or the contents of the *Journal of Economic Dynamics and Control*, you will find out this is the question preoccupying the minds of the researchers.

For me the economic mechanism is *not given*, it is precisely what I am looking for. To use an analogy: they calculate the least expensive way of getting from Amsterdam to Budapest in a particular car, while I try to find out how and from what components a car capable of traveling this route and others can be built. We will see how little the literature on my topic relies on the conceptual and technical instruments offered by the theory of automatic control.

4 On the use of mathematics

I will unfold the message by means of mathematical models and use traditional mathematical tools for their analysis. On the one hand I assume that the reader is familiar with basic concepts and theorems pertaining to matrix calculus and to the theory of ordinary linear differential equations that can be found in any introductory textbook on these topics. I use these concepts and theorems without reference to literature. On the other hand whatever goes beyond that, particularly concepts of control theory, will be explained either in the last three chapters of Part One, or further on as appropriate.

I looked for mathematical instruments suitable for the analysis of a given problem in economics and not — as often happens — the other way round. Even so I am sure that my analysis can be surpassed and rendered more general by the use of more refined mathematics and up-to-date, particularly nonlinear, control theory. I wanted, however, to keep mathematical prerequisites as low as possible in order to make my treatise intelligible to a wider readership. The style chosen freed me from the obligation to separate economic and mathematical analysis. They are interwoven and hopefully will not adversely affect but rather enhance each other.

5 An outline of the book

The following chapters of Part One introduce the basic economic (Chapters 2 and 4) and mathematical (Chapters 5 to 7) concepts needed for an understanding of the remaining parts. Chapter 3 reviews the results of an empirical study reported by KAWASAKI et al. (1982), which provide support for many assumptions underlying my approach.

In Part Two five models present the control of an open Leontief-economy by means of linear controllers using different control signals. They are analysed with reference to stability, viability and structural characteristics.

Part Three is devoted to the study of controllers which are equivalent from the point of view of their operation but represent different organizational structures. The emphasis is on different degrees of centralization concerning information flows and decision making in the control process.

Part Three is concluded by an afterword (Chapter 19). It contrasts the ideas of this book and the related works of others with the Theory of Equilibria with Rationing, the approach closest to the present one, for all its many differences.

Chapter 2

THE ECONOMIC SYSTEM

1 *Enumeration*

At any point in time (t) an abstract economic system consists of the following elements:

— A set $\mathcal{A}$ of *agents* (e.g. households, productive firms, banks, government agencies) which are the subject of the economic activities.

— A set $\mathcal{O}$ of *objects* upon which the economic agents act.

— The natural, historical, social and economic environment $\mathcal{E}$, which is not a part of, but interacts with, the system.

— A set $\mathcal{V}$ of processes which connect elements of sets $\mathcal{A}$, $\mathcal{O}$, and $\mathcal{E}$ and change their state.

When speaking about an economic system the first thing we have in mind is a national economy. Nonetheless, most of the qualifications and methods we use can be applied to systems which are smaller or larger (e.g. an industry, a region).

A particularly interesting task would be to apply the present framework within the theory of industrial organizations to the study of internal working of firms. Although the elaboration of this line falls outside the scope of the present work, I expect that people dealing with that subject would find some useful ideas in it.

2 *Classification of processes and objects*

For a consistent control theory approach, two kinds of economic processes (elements of $\mathcal{V}$) must be distinguished.

Real processes ($\mathcal{V}_r \subset \mathcal{V}$) change the state of physical objects. The most important real processes are: production, storage, transfer of physical objects between agents, consumption (whether for productive or end use). The objects of real processes form the set of *commodities* ($\mathcal{O}_r \subset \mathcal{O}$). The set of real processes consists of the real *activities* of the agents and the *external effects* of the environment. The former depend also on the control processes; the external effects cannot be controlled. The rules which connect the real processes are principally the laws of Nature (or more to the point: technology).

Control processes ($\mathcal{V}_c \subset \mathcal{V}$) change the state of knowledge of the agents and also regulate their behaviour. The objects of these processes ($\mathcal{O}_c \subset \mathcal{O}$) are called *signals*. The most important control processes are: observation of real processes; signal generation and transmission among agents; and decision making (the final signal generation) on real activities. Certain of the signals may come directly from the environment as far as can be observed.

3 Structuring the agents

Since each agent in $\mathcal{A}$ performs both real and control activities, it is convenient not only to split the set of activities and objects into two (real *vs* control) but to consider each agent as consisting of two units: the *real unit* and the *control unit*, which perform the real activities and control respectively. Needless to say, this splitting of an agent into two units is only a conceptual separation, to which an actual separation of the functions may correspond with certain agents (e.g. large firms), but need not in any organized form exist with certain others (e.g. households).

4 Classification of the agents

Finally, to make the dichotomy of the economic system complete, we can even divide the set $\mathcal{A}$ of agents into two subsets: that of *real agents* (or real organizations), $\mathcal{A}_r \subset \mathcal{A}$, whose *main* activities belong to the real processes (like households or productive firms) and that of *control agents* (or control organizations), $\mathcal{A}_c \subset \mathcal{A}$, whose *main* activity lies in information processing and decision making (such as legislative bodies, local authorities, government agencies, trade unions).

This classification of the agents requires some further comment. Firstly there might be borderline agents (e.g. schools) whose classification is ambiguous and will be dependent on the role which they play in a given context.

Secondly, although control organizations perform real activities themselves, mainly in the form of consuming goods and services (paper, manpower, electricity) needed for their basic activity, this real activity is not very important and can in the present abstract treatment be assumed away (in contrast with the control activities of control units of real organizations, which are not negligible and which in fact form the main subject of my study). Incidentally, this assumption is similar to one frequently applied in control engineering, namely that the energy consumption of the regulator is negligible compared to that of the plant which is being regulated.

5 The monetary sphere

Finally a few words are in order about the place in the above dichotomy of *fiduciary goods* (banknotes, accounting money, stocks and bonds), *monetary processes* (emission, exchange, income generation, credit) and *financial organizations* (banks, stockbrokers, tax offices). Since the physical transformation of fiduciary goods is not the focus of economic interest, and hence these cannot belong to the real commodities, they belong to the control sphere by exclusion (in contrast with many other theoretical approaches where money is simply taken as one of the commodities). Even so, it must be borne in mind that the monetary sphere not only plays a particularly important part in the control of economic activities, but differs in many respects from the rest of the control processes and obeys laws which are partly similar to those valid in the real sphere.

However, I do not want to disguise the fact that neither the similarities nor the dissimilarities of the monetary processes to other control processes are sufficiently clarified in the present treatise, but are essentially relegated to future research. On the one hand I admit this to be a shortcoming which must be eliminated in the course of the study of more complex economic control systems. On the other hand this incompleteness will not hinder the basic message of this book where the question of money is deliberately neglected.

The exclusion of monetary processes implies that the flow of goods and services among the agents is not accompanied by a reverse flow of money. There are no incomes and outlays. The acquisition of goods is thus not constrained by the agent's budget. The assumption, however, does not exclude the use of prices in the control process. The formation of prices and their use as signals in the control of the economy can be conceptually separated from the role played by stocks and flows of fiduciary goods, in spite of the fact that prices are expressed in monetary units. But so long as they are only "accounting prices" used in the calculation of control signals without influencing money flows, the formation of prices is a pure control activity.

6 Significance of the dichotomy

After this detour through the monetary sphere we can return to the basic subdivision of the economic system into a real sphere and a control sphere. The *real sphere* subsumes the real processes (real activities of the real organizations plus the external effects from the environment) and the real organizations themselves. The *control sphere* extends to the control units of the real organizations, the control organizations, their activities, the control processes flowing within and between them, including signals coming from the environment.

Consistent application of this dichotomy to the description of economic processes is instrumental to the better understanding of many traditional concepts. It turns out, for example, that while supply usually means a thing existing in the real sphere (goods in store, productive capacity, the manpower in the household), demand is a notion belonging to the control sphere: it is an intention, a decision. According to this logic, excess supply is also a real quantity (e.g. unsold commodities), but the excess demand is a control signal, an unfulfilled disposition to buy. Hence the common practice in many theories of interpreting excess demand as negative excess supply can only lead to conceptual confusion. Another example of this confusion is that of qualifying backlog orders as "negative stocks": although the piling up of inventories is a real process, the generation of orders is a control process.

The distinction between real and control processes has a particular disciplinary power in *model building*. The variables, functions, equations and assumptions can all be classified whether they pertain to the real processes, the control processes or to a connection between a real and a control process. In this respect *Kornai*'s criticism provides quite a convincing argument against such models of the real sphere that produce optimal (shadow) prices as byproducts, without saying anything about the control system within which these prices are formed and take effect.

7 The real sphere

The analytic tool most frequently used in this book is that of adjusting controllers of different types to a certain real sphere and studying their appropriateness and working characteristics. This research strategy admits the choice of a relatively simple description of the real sphere, since this is not after all the focus of my attention.

I have opted for the open static Leontief economy which is characterized by the following restrictions:

a) The list of commodities and the list of agents is constant in time, they neither come into being nor do they die out.

b) The technology is exogenously given (and mainly constant in time) which excludes the study of the mechanism of technological choice.

c) Inputs are proportional to the output produced.

d) End use is exogenous.

Point d) requires three comments. The first is that the content of end use deviates in my world from the usual conceptualization in that it *does not contain the change of stocks* (inventories). Stocks play a particularly important role in the control systems under study, hence it was necessary to separate them from other constituent parts of the end use.

The second remark is more important. By taking end use as an exogenous vector I relegate households to the environment, i.e. renounce the study of consumer behaviour. I focus my attention on the *interrelation among productive firms* and the control of their activities. This omission might pass unnoticed by readers who live in the atmosphere of the research work on mechanisms of socialist economies. But it might be shocking for those who have been reared at the breast of general equilibrium theory. They must be accustomed to the opposite omission: if anything is assumed away, it is precisely the production process. They are interested in the market, and in the market the productive firm is simply a seller and buyer as well as the household. Hereby I once more reveal the guile behind the example in paragraph 1.1; the leading character in my sketch is not the housewife, but the storeman (of materials and finished goods, the production manager, the purchasing agent and the salesman) if we wish only to oversimplify. Nevertheless all the models which follow can be reinterpreted in such a way that each household (or any aggregate of them) is considered a firm producing manpower as output and using subsistence consumer goods as inputs. The end use then contains (among other items) only that portion of consumption which goes beyond the mere subsistence of the manpower. In any case, the peculiarities of the labour market and the particular role of the worker in the production process are not explicitly represented in my models.

8 Dynamics in the real sphere

I deal in my research with changes in economic variables as time passes, with the analysis of the path (trajectory) which an economic system covers, starting from a given initial state to eternity. It is obvious that economic dynamics is the point under discussion.

The attentive reader must have asked already: how on earth can this be? In the preceding paragraph the author told us that the real sphere was represented by a *static* Leontief system, and now he speaks about dynamics? Indeed, I use the static Leontief model in the framework of a dynamic control system, as a part thereof. In effect, the Leontief economy which will appear later is not so static as its name would suggest. Particularly by severing the change of stocks, which is a process in time, from the end use, this model became in a sense dynamic, but not in the same sense as the so-called dynamic Leontief model. In the latter it is the growth of productive capacity and the separation of the investment part of the end use which introduces dynamics; for me it is the balance sheet change of stocks of materials and final products which does this. So there is, indeed, no logical contradiction here, but it would have been misleading to apply the term "dynamic Leontief system", which has long been used in another sense, for the denomination of a real sphere which is made dynamic in a different and narrower sense.

Although it will be clarified along with the mathematical setup what is meant by the term dynamic system, it may not be superfluous to call the reader's attention to yet another terminological hitch. In the language of economic policy and mass communication the term "dynamic" means something forcefully increasing, rapidly developing. This is not the question here, where we apply the term dynamic to an analysis, a model, in which intertemporal effects are examined, and the resulting paths may show any time pattern, whether growth, decrease, fluctuation or stagnation.

9 The control sphere. First comparisons

The last chapter of this book will be devoted to the comparison between the present approach on the one hand and the General Equilibrium Theory, or rather its outgrowth the Theory of Equilibria with Rationing, on the other hand. However, it seems impracticable to expound the basic ideas concerning the control principles applied to my models without drawing a parallel, however sketchy and superficial, with these theories.

10 Dynamics in the control sphere

Equilibrium is a static concept in itself, it is the state of a system where nothing changes, a rest point. Hence the primary concern of equilibrium theory is the existence or nonexistence of an equilibrium state of an economy. Additionally, the stability of the equilibrium state and its efficiency (Pareto optimality) might be analysed. But the process which leads from a state out of equilibrium to an equilibrium state is dissociated from the system and goes on outside real time. In the course of Walrasian auctioning by which the equilibrium prices are formed, nothing happens in the real sphere of the economy. Transactions start only after the equilibrium prices have been set. In the Theory of Equilibria with Rationing the story is similar: both buyers and sellers wait until demands and supplies are contrasted, and agents restricted according to some rationing scheme. These tâtonnement processes run their course in a separate time, or under another interpretation with infinite speed. On the contrary, in my models real processes and control processes run in parallel and affect each other in the same real time. There is no preliminary tâtonnement. Equilibrium theorists are, of course, well aware that the fictitious story of the auctioning (or for that matter rationing) cannot be the last word and have made efforts to get rid of it. I will return to these problems in Chapter 19.

11 *The contents of the control signals*

In the Walrasian theory it is the price called by the auctioneer which serves as the only control signal perceived by the agents. In the Theory of Equilibria with Rationing, prices are (temporarily) fixed and the non-Walrasian equilibrium rationing is reached by using the quantity constraints as perceived by the agents. The "perceived constraints" are already a control signal of a "non-price" or quantity signal character. In my models there are also the quantity signals which play a dominant role. However, my quantity signals consist of such operational notions as the input and output stocks or the stock of backlog orders. In other models conceived in a similar spirit we find other kinds of quantity signals, such as the queueing time in KORNAI - WEIBULL (1978) or a synthetic shortage indicator in KORNAI (1982).

12 *The market*

A market is usually conceptualized as a place where buyers and sellers meet in order to make transactions. There are markets which really exist in this sense (like a stock exchange, or the vegetable market of a small town), but many markets are fictitious, especially in the interfirm transactions in which I am primarily interested.

This remark implies two things. Firstly a description of the markets is only a part of the analysis of an economic mechanism. In my analysis the market will be one of the possible implementations of a control structure.

Secondly one must be cautious in applying the usual assumptions about the working of a market when dealing with less idealized markets.

13 The assumption of "voluntary exchange"

Voluntary exchange means that no buyer (seller) can be forced to buy (sell) more than he wants to (but may be forced to buy or sell less). Although there are markets where this assumption fails (e.g. in many labour markets you will not find a job if you want to work five hours and twenty minutes a day, nor will you be able to buy a ticket for the second act of a play), I follow the usual practice of disregarding these market irregularities and *assume voluntary exchange*.

14 The assumption of the orderly market. The short-side rule

At any point in time on a partial market there might be an unsaleable supply, or unsatisfied demand, or neither, or both, for any particular commodity. If there is neither, i.e. the market is cleared, we speak about "Walrasian equilibrium" (which is not to be confused with the more general equilibrium concepts to be introduced later). If we reject the assumption that market clearing prices develop at an infinite speed, the occurrence of a market in Walrasian equilibrium is highly unlikely.

If in a particular market there might be either unsaleable supply or unsatisfied demand but never both, we say that the market is *orderly*. Behind the assumption of the orderly market we find that whenever there is both a willing buyer and a willing seller on the market they will immediately meet and transact. This assumption implies that in each market either all the buyers or all the sellers are satisfied. Thus each market develops a *short side* and a *long side*. The short side is the supply side if there are unsatisfied buyers (this is then a seller's market) and the demand side if there are unsatisfied suppliers (buyer's market). Hence we say that for an orderly market the *short-side rule* applies.

If we allow the coexistence of both unsaleable supply and unsatisfied demand, then the market is *orderless*. For an orderless market the short-side rule does not apply.

15 *Reasons for the existence of orderless markets*

Why should we assume that a market can be orderless? Is it an exceptional event or a more or less usual feature of some markets? To answer these questions we list a few reasons why a market could operate in an orderless fashion.

a) *Aggregation of markets.* Let us assume that both the green hat market and the grey hat market are orderly, and that there are unsaleable green hats while the grey ones are in short supply. In this case the aggregate "hat market" is orderless. The same argument applies to the aggregation of two locally separated markets of the same commodity. This problem, however, only arises in practical calculations and can be set aside in theoretical considerations, where you deal with as many markets as you like and need not aggregate.

b) *Market frictions.* This is a generic term for phenomena which prevent sellers and buyers from meeting. Such phenomena are, among others, lack of information about the existence of a possible trading partner, delays or errors in the handling of orders, defective bookkeeping of supplies. Some authors consider market frictions as the general cause of the existence of orderless markets, and use the term frictionless or efficient market as synonymous with orderly market.

c) *Discrimination among trading partners.* The traders who appear on the market do not form such a faceless, anonymous mass as economic theories tend to depict them. It is a frequent case that buyers or sellers prefer some trading partners to others. Even in the most free and least regulated market economies there are commodities (e.g. drugs, weapons, spirits) which can be legally sold to licensed buyers only. But the discrimination is mostly deliberate, e.g., racial discrimination on the housing market in some parts of the world. However, from my point of view the relevant discrimination is the preference for an "important" (e.g. military, export) trading partner, or a steady partner, as against minor or stray customers or suppliers. Apparently if a part of the supply is withheld for preferred customers or purchases postponed while waiting for the favourite brand, the short-side rule might be violated and the market becomes orderless.

Among the models that I will present, some can be interpreted as representations of orderly markets and others as of orderless markets.

Chapter 3

THE ECONOMIC CONTROL: PRINCIPLES AND EXPERIENCES

1 *Optimization or survival*

Almost the whole of contemporary theory in mathematical economics is based on the assumption that the agents optimize something (the households their utility functions, the firms their profits for instance), taking technological, financial and market constraints into account. This approach can be (and has been) criticised both from theoretical and from empirical standpoints but I cannot delve here into these disputes. Rather I will simply state that my models do not rely on the assumption that the behaviour of the agents is to be derived from an optimization process. I have followed Kornai's important proposal in AE that the above should be replaced by the less restrictive assumption that the agents make decisions in compliance with certain behavioural rules, whether they are solutions to an optimization problem or not. (This is also a controversial proposition and leads to the "inverse problem of optimal control" which has not yet been analysed in an economic context.)

Assuming optimizing behaviour away, one must face the question of whether qualification weaker than optimality can be applied to tell if a behavioural rule is workable or not. It seems to me that the least artificial criterion can be achieved by assuming that all the agents want to survive. This philosophy leads us to *Viability Theory* to which I will return later.

2 *The "control by norm" principle*

Our favourite housewife decides to bake an apple pie. In doing this she will often revert to "norms", mixing ingredients in proportions which are normal according to her recipe, heating the oven to a normal temperature and baking the pie for a normal length of time. By complying with these norms she will get a normal apple pie. This is an example of an individual (an agent) who uses her own norms in controlling her actions. The source of these norms is her own past experiences (or that of others who taught her to cook), not a command forced upon her by law or public morals.

In her social activities (e.g. inviting guests, working as an employee, acting as a member of a religious community or a sports club) she will control her behaviour according to the set of norms which characterize the social activities in question. These examples all show the importance of norms in the control of human activities in general.

In the case of economic activity norms play no less important a part. The factory manager considers it normal if productive capacities are utilized to, say, 80 percent. An unemployment ratio (5 percent) or a rate of inflation (6 percent) may be considered normal by both economic policy makers and the general public.

The storeman will keep an eye on the stocks of input materials to see whether they are normal for the smooth running of the shop, the sales department observes whether the stock of finished products and/or unfilled backlog orders is at a normal level or not. These are norms which will appear in my models. Empirical support for the relevance of the last two (output stock and backlog of orders) to the control of production and formation of producer's prices will be provided below.

The concept of norm in this interpretation has no connotation of being "desired" or "optimal". I use the word only in the "control by norm" context. This implies that the norm appears as a value to which a control rule (behavioural rule) belongs, which provokes actions counterbalancing any deviation between the actual and the normal value. This is what I call the *principle of control by norm*. In the terminology of control theory *servomechanism* or *regulator* is the technical term used in a similar sense.

3 Other principles

Although in all my models I will apply the control by norm principle, I am not suggesting that this is the only principle which can be applied in the analysis of economic behaviour. It can be contrasted with other control principles. One of them, the *optimization principle*, has already been mentioned. A third may be called the *ward off principle*. When this is applied the control forces are inactive (control variables remain constant) as long as things go well, and an abrupt change takes place in behaviour in the event of trouble, when the consequences of further inactivity are unacceptable. The corresponding control model is called "bing-bang control". I rely on the control by norm principle because I have found it both well suited to the problems I deal with and convenient for mathematical analysis.

4 The dynamics of norms

The concept of a norm is connected to *invariability* over an extended time-span whilst a repeated adjustment to the norm takes place. Even if the adjustment is directed towards a reference path, a time function of a variable, the constant parameters characterizing this path are to be considered as norms rather than the path itself. The norm may appear in a dynamic context as a normal rate of change (e.g. the normal yearly growth rate of nominal wages), a normal ratio of two variables, both changing in time (e.g. the normal rate of profit), a normal time lag (e.g. the gestation period in ship building industry), or in various other forms. After all these reflections, norms are to be considered constant in time.

Even so, we know that norms change in time (and, of course, in space, but this does not concern us now). Firstly, on a historical time scale *norms spring up and die*. The concept of "normal working hours" is the offspring of the industrial revolution, the "normal queueing time for a hospital bed" was born of free medical care. The tithe, also a norm, died out with serfdom. Secondly, *the value of a norm changes with time*. Normal working hours decreased gradually with technical progress, but increased temporarily in wartime. The normal stock of tyres in a bus factory may decrease if the supplier guarantees a shorter time of delivery.

Thus the problem of how and why norms are born, evolve, and die (the dynamics of norms, for short) is an important and interesting issue to study requiring a deep analysis of economic history. And this is a question I do not deal with in the present treatise.

My problem is that of how economic agents control their actions, if once the norms have somehow become fixed. I will try to show that even this shortsighted approach can yield new insights into the working of economic mechanism.

5 Empirical support

In what follows I will rely on the results of an empirical study conducted by *IFO—Institut für Wirtschaftsforschung, München*. The data were analysed and interpreted in a paper by KAWASAKI - McMILLAN - ZIMMERMANN (1982). I use this reference and the reinterpretation of the findings as an empirical support of the following assumptions and starting points of my research.

a) The control by norm principle is actually applied in the behaviour of firms.

b) Different quantity signals, particularly stock and order signals, play a preeminent but not exclusive role in this control.

c) There exist, indeed, many orderless markets.

d) The mere fact that we can start with assumptions which are empirically testable compares favourably with other approaches, which are built on nonmeasurable variables and untestable hypotheses.

Nonetheless, I warn the reader that these tests refer to the premises and not to the conclusions of my analysis, which cannot be expected to be testable due to the abstract setting of my models, where many important effects are omitted from the exercise.

6 The scope of the inquiry

A very simple questionnaire was put to industrial firms in the Federal Republic of Germany during 23 consecutive months from February 1977 to December 1978. Since not all the firms reported each month, the number of data sets varied monthly, but there remained altogether 73,336 observations for each variable. Thus the sample was large enough for statistical evaluation. For 60,800 observations (83%) the firms reported holding both stocks of finished products and unfilled orders and in the remaining 12,536 observations they held a backlog of orders but no output stocks. The questions referred to production lines separately and not to the aggregate output of (usually multi-product) firms.

From these data we can already draw one conclusion. Let us for the moment identify output stocks as indicators of unsold supplies and the backlog of orders as that of unsatisfied demands on the market of the commodities in question. (I will later consider the extent to which this identification is justifiable.) The conclusion is thus that the majority of the markets are orderless and a considerable fraction orderly. Thus assumption 5c is supported by these findings and *it is worth studying both orderly and orderless markets*. (I hope I was cautious enough in not accepting the above proportions at face value. The output of a product line is not necessarily a single commodity but often a variety of similar commodities.)

7 The questions asked

The firms on the panel were asked the following four questions requiring only an answer of +, − or 0 for each product line of the firm:

a) Were the *output stocks* (q) too high (−), satisfactory (0), or too low (+) in the previous month?

b) Was the *stock of backlog orders* (k) too high (+), satisfactory (0), or too low (−) in the previous month? (To wit, the signs were chosen so that + indicated the direction of excess demand for both questions.)

c) Did the *quantity of production* (r) increase (+), remain constant (0), or decrease (−) in the current month as compared to the previous one?

d) The same question concerning the *price* (p) of the product.

Although the questions themselves fail to put it into words, they imply the existence of a level (or at least a range) of output stocks and unfilled orders considered sufficiently "normal" by the firms as to serve as the basis for the value judgements "too low" and "too high". Although in their analysis the authors refer to these norms (in my terminology) as optimum levels this feature is neither needed nor made use of. (An example of how the optimization principle creeps into the thinking of economists even if it is uncalled for.) In any case, the above questions indicate that the analysis of the data is highly relevant to my studies.

8 The methods and the data

In the paper referred to above the authors process and evaluate the data by mathematical-statistical methods. I have no reason to deal with the details of their method here, and the interested reader should refer to the original publication. The method is a combination of the multivariate conditional logit model developed by *Nerlove* and *Press* with the γ coefficient method of *Goodman* and *Kruskal* extended by the authors. The γ coefficient is an indicator analogous to the usual correlation coefficient in expressing the closeness of the connection between two variables and taking values from the interval $[-1, +1]$. The difference is that the strength of the connection is not measured according to the size of the deviations (which has not been observed), but to the frequency of their occurrence in one or the other direction.

The following γ coefficients were calculated:

$\gamma(r,q)$ = dependence of production on output stocks

$\gamma(p,q)$ = dependence of price on output stocks

$\gamma(r,k)$ = dependence of production on stock of unfilled orders

$\gamma(p,k)$ = dependence of prices on stock of unfilled orders.

In addition, asymptotic t-ratios were calculated and marked as to whether the coefficient was significant at the 5 percent level. φ^2 indicators were calculated to evaluate the residual effects of variables not represented in the model, and χ^2 tests to check the null hypothesis of no dependence. This later resulted in the rejection of the null hypothesis in every case.

9 Selected numerical results

The results below (Table 1) are taken or calculated from Tables 1, 2 and 3 *loco citato*.

Table 1

		Value of the γ coefficient	Asymptotic t-ratio	Proportion (%) of *months industries* where the coefficient was significant	
Firms with output and order stocks	$\gamma(p,q)$	0.199	6.95	35	11
	$\gamma(r,q)$	0.249	14.09	91	67
	$\gamma(p,k)$	0.324	10.96	78	33
	$\gamma(r,k)$	0.475	29.73	100	78
Firms without output stocks	$\gamma(p,k)$	0.475	7.59	.	.
	$\gamma(r,k)$	0.496	7.98	.	.

10 What do the numerical results show?

The numerical results selected for the table above permit the following conclusions to be drawn:

(i) The γ coefficients are all *significant* on the aggregate level (both across time and industries), but they are insignificant in certain months and for certain industries. The frequency of insignificance is different for the different γ's. This shows that all four supposed dependencies take effect in the real world.

(ii) All the γ's are *positive*, which shows that the reaction of the firms is mostly in the right direction, i.e. they try to counterbalance the deviations from the norms.

(iii) *Quantity adjustment* (change of the production level) is more marked than price adjustment in the short run.

(iv) Too high or too low *stock of backlog orders* affects the decisions of the firms more forcefully than do the output stocks.

Additionally the authors report that "in about 60 percent of the observations firms were in disequilibrium with respect to inventories, unfilled orders, or both" which indicates the importance of disequilibrium adjustments. It is a pity that they fail to disclose more either about the apprehension of disequilibrium or about the change of production and prices in the cases of equilibrium, which run to 40 percent of the total. This would have told more about the effects of the neglected factors. We are told that the unpublished φ^2 coefficients reveal that unobserved factors (like cost changes or lasting disequilibrium) affect the prices more than production levels.

I will return repeatedly to these findings in the course of the analysis of my models in order to confront theoretical constructs with reality.

Chapter 4

CONTROL STRUCTURES:
VEGETATIVE FUNCTIONING AND COORDINATION

1 *Insufficiency of the binary concept of centralization vs decentralization*

In this chapter I will try to show that the binary distinction between the control structures of economic processes — namely that there are on the one hand centralized processes and systems and on the other hand decentralized ones — is not satisfactory. I propose a new and more complete classification.

The problem itself is "classical" in the sense that from the very beginning of economics as a science, this has remained one of the primary — and perhaps the least clearly expressed — questions under study. This was the problem pried into by *Adam Smith* when bringing up the "invisible hand". *Walras* speaks about the no less extraneous "auctioneer" instead, while nowadays in disequilibrium theory the "rationing agent" plays an equally mystical role.

The problem of centralization stood in the forefront of the theoretical controversy on socialist economy in the thirties as well as in the reform debates from the late fifties onwards. The misunderstanding of this problem lies behind the miscarried "economic reforms" restricted to organizational restructuring while leaving operational modes unchanged. In Part Three I hope to show just how controllers of different centralization patterns may be (in a given sense) equivalent in their operation.

The explanation of the classification schema starts with an elaborate introduction of certain concepts. This is presented here in a purely verbal way, while an embryonic formalization will be attempted in Part Three.

2 Dissection of control processes. The transfer element

Let us imagine a relatively simple (single-loop) control circuit, i.e. one where there is no feedback within the controller itself. (The issue of control loops and their classification will be discussed in the next chapter.) In such a case the control process is composed of consecutive *sub-processes* and parallel *partial processes*. For example, each firm observes separately the inventory of its products, these are the partial processes of the output stock measurement process. Taken together they form the sub-process of stock taking which is followed, e.g., by the sub-process of price setting. This dissection is illustrated in Figure 1. The control process is composed here of four sub-processes (I-IV) and from among them II and III are each divided into three partial processes. Every process which is not further divided into sub- or partial processes is called an *elementary process*; each elementary process is performed by a single *transfer element* of the controller. The transfer element is thus that part of the (usually complex) controller which is indivisible or at least undivided in the given analysis.

3 The functioning of the transfer element

In the functioning of every transfer element three consecutive *acts* take place. In the description of the acts the keyword is *signal*, the transfer element is a producer and consumer of signals. The three acts are:

Signal reception: The transfer element receives signal (information) directly from the real process or the environment or another transfer element. These are termed the *input signals* of the element.

Signal generation: The transfer element transforms, combines and stores the received signals and hereby generates new signals. The rule of signal generation is called the *response function* of the element.

Signal emission: The transfer element transmits the generated signals, the *output signals*, either to another transfer element or to a manipulating organ intervening directly into the real process.

If in the complex control process the signal is both emitted and received by a transfer element, then the signal emission and signal reception are in fact two projections of a single act, the *signal*

Figure 1

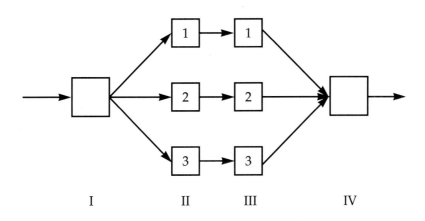

I II III IV

Dissection of a controller

transmission. For the sake of simpler description we will take such an act into account always from the point of view of the element emitting the signal. This simplification can, however, be applied neither to signals coming directly from the real processes or the environment, nor to those going to a manipulating organ. They will soon be dealt with separately.

The elementary control processes will now be classified according to two criteria both relating to the kinds of agents (organizations) taking part in the process:

a) What kind of organization generates the signal?

b) Between what kind of organizations is the signal transmitted?

4 Classification of signal generation

Three kinds of elementary control processes can be distinguished from the point of view of the organizations taking part in the signal generation:

Uncoordinated: the signal is generated by the control unit of a real organization.

Interactive: the signal is generated jointly by the control units of several real organizations.

Centralized: the signal is generated by a control organization (perhaps jointly with other control or real organizations).

The uncoordinated and interactive processes are referred to by a common name — *decentralized* — while the interactive and centralized processes are referred to as *coordinated* processes.

Besides replacing the traditional centralized-decentralized dichotomy by the above trichotomy, I will also supplement it with another dimension, that which refers to the signal transmission.

5 Classification of signal transmission

From the point of view of signal emission and transmission between organizations we can also divide the processes into three classes:

Non-communicative: the signal does not leave the organization from whence it originates.

Transactionally communicative: the sender and the addressee are different real organizations and the signal refers to an actual or potential real process (usually a product transfer) between them (e.g. an order, a price quotation, a bill).

Non-transactionally communicative: any other signal transmission, e.g. between more than two real organizations, between a real and a control organization, two or more control organizations.

6 Combination of signal generation and transmission

We can combine the two dimensions of signal generation and transmission into a table of examples. (Table 2)

The table contains two empty boxes, these generation-transmission combinations cannot occur. An interactive signal is generated jointly by several real organizations, and cannot go without communication between them. In centralized signal generation a control organization takes part; the subsequent signal transmission cannot be transactional.

7 Vegetative functioning of a transfer element

The functioning of a transfer element is said to be *vegetative* if the signal generation is uncoordinated, and the transmission either noncommunicative or transactionally communicative. The term *vegetative* makes allusion to the lower autonomous functioning of a nervous system and suggests the incapability of this kind of control to bring on development. (I here define vegetative functioning of a *transfer element* only. The important concept of the vegetative functioning of a *system* will be explained shortly.)

Table 2

SIGNAL TRANSMISSION	SIGNAL GENERATION		
	Uncoordinated	*Coordinated* Interactive	Centralized
Non-communicative	Storeman takes inventory and reports it to production manager	— —	Emission of an intra-office memorandum in the Ministry Agriculture
Transactionally communicative	Buyer decides on order and sends it to the supplier	Prices are set in a bargaining process among buyers and sellers	— —
Non-transactionally communicative	Firm reports payroll data to the Statistical Office	Chemical firms agree to build a joint sewage farm and report to the Town Council	Planning Office sets export target for a firm

8 The start and the end of the control process

The foregoing discussion of the transfer elements is valid only to such as are connected with control processes both on the input and the output side. The total control process, however, is also linked with the real processes, this link being provided by special organs.

At the beginning of the control process we find a special transfer element called *sensor*, which makes *observation* or *measurement* of variables of the real processes, the controller receiving the output signal of the sensor.

At the end of the control process we find another organ, the *manipulator* or *effector*, which intervenes with the real processes. The control process ends with the *decision on intervention*, this decision being conveyed to the real processes via manipulating variables.

For the sake of simplifying the subsequent discussion I make a twin assumption on the sensor and on the manipulator, to be applied throughout the book without further comment.

9 *Assumption: The sensor is introspective* The measurement or observation process as well as the sensor is said to be introspective, if any real process can be directly observed only by a sensor belonging to the control unit of a real organization in whose real unit the real process has been carried out. A control organization or an outside real organization cannot learn about the value of a real variable unless told by an involved party. (In the opposite case we should speak about spying, this being what is excluded by the assumption.)

10 *Assumption: The manipulator is introactive* The manipulation of a real process as well as the manipulator is said to be introactive, if any (final) decision on a manipulating variable is made by the control unit of a real organization in whose real unit the decision will be implemented. A control organization or an outside real organization cannot influence the final decision on the value of a manipulating variable unless it does so via the control unit of an involved party. (For example, a commodity transfer by theft or confiscation is excluded by this assumption.)

These assumptions facilitate the tracing of the effect of a signal from the measurement up to the final decision and implementation.

11 *The identity transfer element*

A transfer element is called an identity transfer element if its input signal equals the output signal and if there is no delay between input and output. Sensors and manipulators will mostly be assumed to be identity transfer elements. This assumption implies error-free transfer between actual and measured values, or between final decision and implementation, respectively.

12 *The complex controller*

Having given a taxonomy of *elementary* control processes and the carrying transfer elements I will now turn to the structure of a *complex* control process and controller, always from the point of view of degrees of centralization.

It must be clear that transfer elements representing signal generation and transmission of different degrees of centralization can be combined in various ways to form complex processes. Thus complex controllers cannot be arranged in a one or two dimensional scheme. It is generally not possible to tell of two complex controllers which is the more or the less centralized.

Nevertheless I have arbitrarily chosen five typical complex control processes, such that they can be arranged in a sequence of increasing centralization. Many of them will be represented by the models of Part Two and Three.

13 *First stage: Vegetative non-communicative control*

This controller consists of transfer elements with uncoordinated signal generation and non-communicative transmission. Hence in this controller the whole control function is performed by the control units of the real organizations, no control organization can exist at this stage. Furthermore there is no transmission of information between the participants of this system. A surprising theorem will show that under appropriate conditions such a system can be viable.

The question arises as to whether in this case there is any connection between the real organizations or whether the whole system disintegrates into its constituent parts. The answer is that there is such a connection via the real activities. In this way the control units — although the measurement is introspective and there is no exchange of information — gather signals on the activities of others by the observation of their own. The most frequent form is where the seller knows the real activity, the purchase by the buyer, even if decided by the latter, and conversely.

14 *Second stage: Vegetative control with transactional communication*

This stage accords with the first one in that the signal generation is uncoordinated, but differs in information structure. Here we have direct (transactional) communication between pairs of real organizations. Even so, there is no control organization at this stage. Models representing this stage will differ according to the contents of transmitted signals, we will meet both quantity (order) and price signal based systems.

15 *Third stage: Interactive control*

At this stage there is still no control organization involved, but there appears an elementary process where several real organizations generate some signal jointly. The process is thus coordinated but the signal need not have an institutional sender. Such a role is played by the market price in the case of atomistic competition. One cannot tell who set the price; it is the result of an anonymous social process with many actors. A model in Part Three can be interpreted this way.

16 *Fourth stage: Partially centralized control*

This stage is characterized by the appearance of one or more control organizations which control certain sub- or partial processes, but none of which keeps all the control processes in hand. The control units of the real organizations are usually not restricted to the observation of the real processes, but participate in the control of other partial processes as well.

In partially centralized control the term "partial" may mean different things:

a) Some real organizations belong to the scope of centralized control, others do not (e.g. centrally controlled industry and interactive agricultural markets).

b) Among consecutive partial processes we find both centralized ones (e.g. centralized price setting) and uncoordinated or interactive processes (e.g. buyers decide on purchases in isolation).

c) Some groups of signals (e.g. the allocation of some basic raw materials) are centrally controlled, while others are not.

This enumeration is not exhaustive and many other kinds of partially centralized controllers may certainly be constructed. Obviously this stage is the one closest to reality.

There might be several partial control organizations in a system, in which case we speak of *polycentric control*. Such control organizations usually form a hierarchical system. In this book the particular problems of hierarchical systems are not dealt with. Among the rich literature on this subject the reader may consult the book by MESAROVIC - MACKO - TAKAHARA (1970).

17 *Fifth stage: Monocentric control*

Here we find a "supercentre" which performs the whole control activity from collecting the data from the real organizations up to the final decisions on real activities. Since with monocentric control all the necessary information is concentrated in one pair of hands it is clear that if a control problem can be solved at all, it can also be solved by monocentric control (disregarding, of course, problems pertaining to reliability, speed and expenditure as well as the efficiency of such an arrangement). The opposite statement is false, the less information the individual decision makers possess, the harder it is to construct a decentralized well-functioning system from them. Monocentric control will not be modelled in this book.

18 *A comparison of the stages*

The above line of reasoning should have convinced the reader that the "rough" division of the control systems into centralized and decentralized ones is not satisfactory. According to this rough division, the first and second stages qualify as decentralized, the fourth and fifth as centralized.

Actually the difference between the first and the second stage is essential: in the first stage the economic agents are connected only by the exchange of commodities, in the second by an extended information network as well; the agents send signals to each other, although their activity remains uncoordinated. I think that the modelling of the first and second stages — not previously dealt with and to which several colleagues of mine have contributed — is a true achievement.

The classification which (under the heading of "centralization") blurs the difference between the fourth and fifth stage is also too vague. The importance of this distinction is well known from real life. The control activities of the various control organizations (such as Planning Office, Price Authority, Central Bank, local authorities) may not be sufficiently coordinated. The theory has not yet made much progress with these problems. My studies are also restricted to comparisons with lower-stage controls in Part Three.

The third stage, interactive control, seems to be the most controversial. The proposition that the market operates as an information centre is not new at all. Although I will also illustrate the working of such a market I cannot get beyond the unfortunate tradition of describing the working of the market as a "black box". This is the same as that which was done in the whole neoclassical literature when describing the dynamics of price formation on the market, from ARROW - HURWICZ (1960) to ALLAIS (1981), who perhaps penetrated furthest in this area. I am attributing to this sort of description the fact that I cannot formally distinguish between price formation on the market *vs* in a Price Office; both are collectors of quantity signals, transforming them into price signals. Hence the difference between the third and fourth stage remains obscure. The outcome is the same whether we are dealing with a Price Office of *Lange*, or an interactive market Stage Three. Their separation must be left to other types of models, this being perhaps the most interesting open question in economic theory to emerge from my approach. I can only hope that the above conceptual framework, with its sharper formulation of the problem, clearer distinction of stages and components of centralization, may contribute to future progress in the theory of centralization.

Chapter 5

SYSTEMS CONTROL

1 *Control systems*

Hitherto we have followed a reasoning which — although influenced by the mathematical control theory paradigm we are going to apply — derived its conceptualization from the economic systems under study. Now we begin to approach our subject from the other direction, that of control theory, which has developed as a mathematical discipline in the service of control engineering. Thus in the following chapters of Part One I will disregard the economic contents and concentrate on a formal presentation of a control system. Both the topics selected and the ways of presentation are, however, fitted to the needs of the subsequent analysis of special economic models. Therefore at least a superficial glance over the next three chapters is recommended for those readers who are familiar with mathematical control theory.

The most common problem of control theory is the following: *How should we intervene in a process evolving over time in order that the process would meet certain requirements*?

The processes we wish to direct to a desired path (they are economic processes in our case, and physical, chemical or biological processes in natural sciences and engineering applications) are called *controlled processes* or *controlled subsystems* (in engineering literature the term "plant" is frequently used). The intervention is provided by the *controller*. It is usually assumed that the laws of the controlled process are given and cannot be changed at will, but we have some freedom to adapt the controller to our purposes and influence its structure and means of operation. For instance, the car driver takes the construction and operating laws of the car as given, but it is in his power to convey his intentions by means of control devices (accelerator, brake, steering wheel), and in this way to direct and control the time-path of the car. In my models the controlled process consists of the change of stocks and backlog orders on the effect of production, material consumption and transfer of commodities, and the intervention consists of somehow influencing the amounts produced, purchased, sold or ordered. The controlled subsystem and the controller taken together form a *control system*.

2 *Signals*

The desiderata conveyed to the controller are called *command signals* or *reference inputs* which may be constant or may vary with time. The signals generated by the controller and influencing the course of the controlled process are called *manipulating variables*. Besides the manipulating variables transmitted from the controller, the controlled processes are also affected by environmental factors which we cannot influence. These are called *external effects,* and in control theory the terms "load" and "disturbance" are also in use. At the end of the controlled process the *controlled variables* appear. When driving a car the position of the accelerator is a command signal, the revolution per minute of the engine is a manipulating variable and the speed of the car is a controlled variable. The steepness of the road is an external effect. In my models I have for example stock norms as command signals, amounts of commodities produced and transferred as manipulating variables, stocks as controlled variables and final use as external effect.

The external effects may be directly observed but usually are not. In the latter case their presence is only inferred indirectly from observation of the controlled variables. For example, the car driver can sometimes perceive a change in inclination with his eyes, while in other instances he only infers that change from the increased or decreased speed of the car.

3 *Open-loop control*

All this being said, the simplest schema for a control system is shown in Figure 2. In what follows this schema will be elaborated with more and more details.

Figure 2

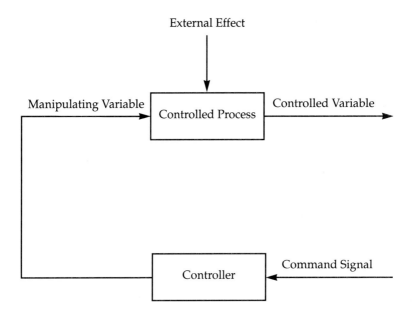

Open-loop control

In the figure the controlled variable depends on the manipulating variable (and on any external effect), but the converse does not obtain. The manipulating variable is influenced only by the command signal but not by the value of the controlled variable. In order to distinguish it from what follows, this kind of control is called *open-loop*. Open-loop control is infrequent in the real world, and this schema is mostly used to approximate reality if we are ignorant or unconcerned as to how the controller reacts to the characteristics of the controlled process.

4 Feedback control

In the majority of control processes at least a part of the controlled variables is observed (measurement). The information obtained (selected, processed and so on) is then compared with the command (reference) signal. The difference between the actual and reference value of the controlled variable is called the *actuating signal* which puts the controller into operation. This arrangement is termed *closed loop* or *feedback* control and the system provided with feedback is a *control circuit*. Its schema is shown in Figure 3.

A classical example of feedback control is the thermostat. The controlled variable is the temperature of the thermostat, this is measured and compared with the desired temperature, the command signal. The controller changes the amount of heat flowing into the thermostat; the manipulation is the transfer of an amount of heat. The external effect is the quantity of heat flowing from (or to) the environment which cannot and need not be directly observed.

5 Classification of controls

We have already distinguished open-loop control from feedback control, and in the sequel other criteria will be discussed qualifying different kinds of control systems. Namely we will make the following distinctions.

a) *Continuously* and *intermittently* operating systems

b) *Single-loop* and *multi-loop* systems

Figure 3

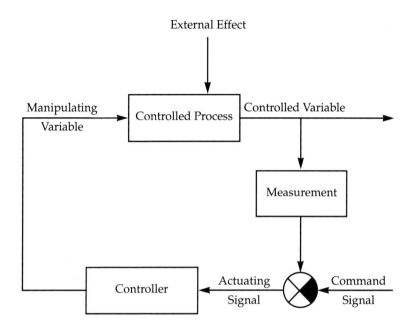

Feedback (closed-loop) control

c) *Time-invariant* and *time-variant*, *linear* and *nonlinear* systems

d) *Deterministic* and *stochastic* systems

e) *Externally commanded* and *higher order* systems

f) *Anticipatory* and *non-anticipatory* systems

6 Continuously and intermittently operating systems

While in continuously operating systems the controller may change the manipulating variable at any time, in intermittently operating systems this change is restricted to selected points in time. Between two such discrete points in time the manipulating variable remains constant. In the description of continuously operating systems time (t) is a continuous variable which may take any real value. We operate our systems, however, from some initial point in time (e.g. zero time) and thus usually restrict ourselves to non-negative t values. In this book I discuss continuously operating systems which can conveniently be described by differential equations.

7 Single-loop and multi-loop systems

Single-loop systems are characterized by the feature that if the process is broken at a single point it becomes open-loop control. Multi-loop systems do not possess this feature; they comprise several interwoven circuits, and an element which plays the role of controller in one loop may appear as a controlled element in another loop.

In this context let me make a terminological digression. In Chapter 2, I stressed the importance of distinguishing the real sphere and the control sphere, real processes and control processes from the point of view of the economic analysis. Here, when discussing the control systems in a more general context, I distinguish controlled processes (subsystems) and control processes (controllers). The question arises as to whether these two kinds of conceptual division coincide or how much they differ. In the simplest models of an economic system the controlled processes are chosen from among the real processes of the economy, the control processes from the control sphere of the economy. But in a real life economy, or even in a more complicated multi-loop model of it, some processes of the economic control sphere may appear as controlled processes (e.g. price control, the control of money circulation). In some of my models the piling up of backlog orders (which is clearly a process evolving in the control sphere of the economy) appears as a part of the controlled subsystem. Thus we are justified in differentiating between the two (economic *vs* control theoretical) kinds of classification.

8 Time invariant and time-variant, linear and nonlinear systems

A system is time-invariant if its parameters are independent of time (in both controlled and control processes), and is time-variant if some of them are exogenously determined functions of time. I will mostly discuss time-invariant systems for the sake of simpler presentation, but many of the results can easily be extended to time-variant systems subject to restrictions on the variable parameters.

A system is called *linear* if, in the mathematical formulation of both the controlled and the control processes, there occur only linear equations (algebraic equations, differential equations, difference equations) and is called nonlinear otherwise. I will deal only with linear systems.

The two preceding "definitions" are tautological and very inaccurate. I fell back on them in order to avoid here the many mathematical concepts which are needed for an exact definition of the above properties. The interested reader may consult Chapter 3 in ZADEH - DESOER (1963), or any other textbook on linear systems or control theory.

I feel obliged to explain why I think it is a venial sin to confine myself to linear models, being aware that real world economies are nonlinear. Is this not an inadmissible simplification? Without denying that a more general elaboration of the subject would have been more convincing and might yield more significant results, the answer is no. My fundamental purpose is to analyse the viability of certain economic control structures *qualitatively*. Hence if I can construct viable systems in linear form, I will have made my point. Even so, I admit that in one respect this comfortable attire proved to be a strait-jacket. That is to say, a number of economic relations could have been conveniently formulated in inequalities rather than equations. But (even a linear) inequality makes an otherwise linear system nonlinear. Thus I solemnly sacrificed inequalities on the altar of God Linearity (and re-smuggled some of them in the guise of Viability Theory).

9 Deterministic and stochastic systems

An output signal of the controlled subsystem or the controller is deterministic if it depends uniquely on the state and input signal of the subsystem. If both subsystems emit only deterministic signals we speak about a deterministic system. In the other case, if at least one output signal can take random values, we call the system stochastic. Nothing is said here about the nature of the input signals (command signals, external effects). The input signals may be stochastic without affecting the deterministic nature of the system, e.g. the thermostat works deterministically if it provides a definite amount of heat in response to a definite (inside) temperature. The fact that the environment induces random heat losses, or that the person who sets the desired temperature might make random changes, does not render the system stochastic. I will not discuss stochastic systems in this book.

The reasons for the appearance of stochastic processes in a control system are manifold. They may be present in the controlled subsystem (e.g. breakdown of a machine, uneven performance of the workers), in the sensor (measurement error), in the controller (defective signal generation, noise in the transmission channel, erratic decision making) or in the manipulator (misimplementation of a decision). These interior disturbances are all excluded by my assuming stochastic processes away.

10 Externally commanded and higher order systems

In the simplest control models the command (reference) signal is given as a constant or an exogenous function of time, and the structure, the parameter set of the controller is also fixed. In higher order (e.g. biological and social, among them economic) control systems this simplistic approach is largely inadequate for the representation of a whole, complex entity, but may serve as an approximation to the description of a small set of processes torn from many contexts in which they appear. The higher order functioning of a control system would also include some or all of the following aspects:

a) *Target modification and self-command.* This requires a system in which the command signal is not exogenous but is generated within the system. (Cf. 3.4 on the formation of norms.)

b) *Learning.* This function implies the recording of past experiences on the performance of the controller (memory), changing the parameters of the controller and adjusting them in view of lasting variations in the environment or in the interior functioning (e.g. aging) of the system. Such controls are often called *adaptive*, but the meaning of this attributive is not fixed in the literature.

c) *Self-organization.* A self-organizing system is capable of changing the structure of the controller, the component elements of both the controlled and the control subsystems, and the connections among them. Needless to say, most systems of human society display such functions, and this is perhaps the characteristic difference between biological and social systems.

It must be quite clear by now what a breakthrough it would be to formalize adequately the higher order functioning of socio-economic systems. As long as our science is not sufficiently developed to encompass these aspects in a well-conceptualized and formalized framework, we are only preparing ourselves for the task of establishing a comprehensive theory. This being the state of the art, I confine myself not to go beyond the simplest problems.

11 *Anticipatory and non-anticipatory systems*

In any law governing a natural system it is forbidden to allow present state changes to depend on *future* states: past states perhaps (in systems with memory), present state certainly, but never future states.

This postulate, causality, is deeply rooted in all sciences. Therefore anticipatory systems, which would contain just the reversal of the causality chain, have always been prohibited in physics and engineering as "non-realizable" systems.

As regards biology and social sciences, however, the case is somewhat different. Such systems may contain a model (not necessarily a mathematical model) of themselves and/or their environment which produces predictions on future states. Such predicted future states can then be used as arguments in decisions on present actions, without violating causality. In economics anticipatory behaviour in this sense is abundant. Price speculation, investment decisions, planning (on either the micro or macro level) cannot be explained at all without anticipation.

In this book, however, I deal with non-anticipatory systems only. I am aware again that this restriction bars the modeling of many basic economic processes and lends a strongly mechanical tone to my reasoning. But I think that the study of economic control systems should have started with non-anticipatory systems, if for no other reason than the separation of non-anticipatory constituents of economic behaviour from the higher order anticipatory modes. The study of the latter, which will imply the modelling of (predictive or planning) models within the model of the economic process, is left to future research.

Chapter 6

STATE EQUATIONS OF THE TRANSFER ELEMENT AND THE CONTROL CIRCUIT

1 Methods

I am unable to say more concerning the theoretical background and problems of my treatise without using mathematical formalism. The further discussion requires the tools of mathematical control theory, to which I now turn my attention.

There are several methods available for the formal description of control systems. Among the *graphical methods*, block diagrams and signal flow diagrams are in use and I will illustrate my models with block diagrams by and by. However, for a complicated, multivariable system graphical representation loses its main virtue of lucidity.

A dynamic, continuously operating system can be most favourably formulated by *differential equations*. Systems of differential equations can frequently be reformulated into *state equation* form. This is especially suited to linear systems where matrix algebra offers powerful analytic tools as does the Laplace transformation for time-invariant systems. I count on the reader's versatility with regard to matrix operations.

In the present chapter I introduce only the general differential equation and the state equation forms, while the Laplace-transform will only be introduced later at the place of application (Chapter 15). I will start with different forms of a transfer element and will return to the control circuit afterwards.

2 General differential equation form of a linear transfer element

A transfer element is characterized by the vector v of its *input signals*, the vector y of its *output signals* and the interrelation between them.

We do not care here where the input signals come from (output of another transfer element, sensor, external effect, command) nor where they go.

A linear transfer element can be represented by the vector differential equation 3, where differentiability (with respect to time) of the appropriate order is tacitly assumed, as it will be throughout the sequel.

3
$$A_k \frac{d^k y}{dt^k} + \ldots + A_2 \ddot{y} + A_1 \dot{y} + A_0 y = B_m \frac{d^m v}{dt^m} + \ldots$$
$$+ B_2 \ddot{v} + B_1 \dot{v} + B_0 v,$$

where

$v = v(t)$ is the μ-vector of input signals

$y = y(t)$ is the υ-vector of output signals

$A_i = A_i(t)$ (i = 0, 1,..., k) are υ×υ square matrices

$B_j = B_j(t)$ (j = 0, 1,..., m) are υ×μ matrices.

k is called the *order* of the differential equation 3, if A_k is a non-zero matrix in $t \in [t_0, +\infty]$. Usually the assumption that A_k is *nonsingular* for all $t \in [t_0, +\infty]$ is also made. The matrices A_i, B_j are assumed to be continuous in $[t_0, +\infty]$. Should all the matrices A_i (i = 0, 1,..., k) B_j (j = 0, 1,..., m) be independent of time, the system becomes time-invariant. In this case the notation used for these matrices changes to A_i, B_j.

4 PID elements

Let us specify 3 in the following way:

$$k = 1, \quad m = 2,$$
$$A_1 = E, \quad A_0 = 0,$$
$$B_j = B_j \quad (j = 0, 1, 2)$$

so that we obtain the equation:

5
$$\dot{y} = B_0 v + B_1 \dot{v} + B_2 \ddot{v},$$

and integrate this equation with respect to t, disregarding the additive constant of integration.

6
$$y = B_0 \int v dt + B_1 v + B_2 \dot{v}.$$

The transfer element corresponding to this specification is called a PID element, and the terms on the right-hand side are called integrating (I), proportional (P) and differentiating (D) terms in the order they are written in 6. If only two of the three terms appear we speak about a PI, PD or ID element. The simple P element is *static* (i.e. represented by an algebraic rather than a differential equation; this criterion applies also to nonlinear elements).

7 State equation of a linear transfer element

A time-variant linear transfer element can also be described in the following state equation form:

8
$$\dot{x} = Px + Ru$$

9
$$y = Sx$$

where

x is the *state vector*

u is the *control vector* (or *input vector*)

y is the *output vector*

P is the *system matrix* (square)

R is the *input matrix*

S is the *output matrix*.

The matrices P, R, S are again continuous in $[t_0, +\infty]$ by assumption and so is u. If the time dependent matrices P, R, S are substituted by the constant matrices **P, R, S** then 8-9 represents a *time-invariant* transfer element.

We have not specified the dimension of the vectors x, u and y above. But it is to be noted that they may be different. The corresponding real Euclidean spaces are called *state space, control* (or *input*) *space* and *output space* respectively.

10 Transformation of the general differential equation form to state equation form

The form **3** may be brought into the state equation form **8-9** in several ways. Different state equation representations of **3** can be achieved by different specifications of the state vector x. If A_n in **3** is nonsingular, then the following transformation into a form, which is called *phase space* representation, is both simple and well suited to physical interpretation. Let us denote

$$x := \begin{bmatrix} y \\ \dot{y} \\ \vdots \\ \dfrac{d^{k-1}y}{dt^{k-1}} \end{bmatrix}, \qquad u := B_0 v + B_1 \dot{v} + \ldots + B_m \dfrac{d^m v}{dt^m}.$$

Then equation **3** can be replaced by the following pair:

11
$$\dot{x} = \begin{bmatrix} 0 & E & 0 & \ldots & 0 \\ 0 & 0 & E & \ldots & 0 \\ \cdot & \cdot & \cdot & & \cdot \\ \cdot & \cdot & \cdot & & \cdot \\ \cdot & \cdot & \cdot & & \cdot \\ 0 & 0 & 0 & \ldots & E \\ -A_k^{-1}A_0 & -A_k^{-1}A_1 & -A_k^{-1}A_2 & \ldots & -A_k^{-1}A_{k-1} \end{bmatrix} x + \begin{bmatrix} 0 \\ 0 \\ \cdot \\ \cdot \\ \cdot \\ 0 \\ A_k^{-1} \end{bmatrix} u$$

12
$$y = [E, 0, 0, \ldots, 0]x.$$

This way the system **11-12** with an obvious notation corresponds to the form **8-9**.

13 The case when the state equals the output

It is a particular but frequent occurrence that the state vector x coincides with the output vector y, i.e. we have $S = E$ for all t and **9** reduces to $y = x$, which is then omitted.

14 *The solution of the state equation: the time-variant case*

The state equation **8** can be explicitly solved for x, the trajectory of x depends on the initial state at the point of time t_0

$$x(t_0) = x_0,$$

as well as on the control input trajectory $\tilde{u}(\tau)$, for $t_0 \leq \tau \leq t$. (The tilde on u is a reminder that x is dependent on the whole past trajectory of the input u.)

15
$$x(t) = x(t,t_0,x_0,\tilde{u}) = \phi(t,t_0)x_0 + \int_{t_0}^{t} \phi(t,\tau)R(\tau)u(\tau)d\tau,$$

where the square matrix $\phi(t,t_0)$ is called the *state-transition* matrix. It is defined as the solution of the matrix differential equation:

16
$$\dot{\phi}(t,t_0) = P(t)\phi(t,t_0), \qquad \phi(t_0,t_0) = E.$$

17 *The state transition matrix in time-invariant case*

For time-invariant systems ($P(t) = P, \forall t \geq t_0$) **16** can be easily solved:

18
$$\phi(t,t_0) = \exp[P(t - t_0)].$$

Since in the time invariant case we can put $t_0 = 0$ without loss of generality, **18** can be simplified to

19
$$\phi(t) := \phi(t,0) = \exp(Pt).$$

20 *The solution of the state equation: the time-invariant case*

In view of **19** the solution **15** of the state equation takes the following form ($t_0 = 0$, $x(0) = x_0$, $P(t) = P$, $R(t) = R$):

21
$$x(t,x_0,\tilde{u}) = \exp(Pt)x_0 + \int_0^t \exp[P(t-\tau)]Ru(\tau)d\tau.$$

If — additionally — the input signal is constant ($u(t) = u, \forall t \geq 0$) and

P is nonsingular, then the integration on the right-hand side of **21** can be carried out so that we obtain the constant input solution:

22 $$x(t,x_0,u) = \exp(Pt)(x_0 + P^{-1}Ru) - P^{-1}Ru.$$

We will now turn our attention to certain particular states.

23 The zero-input equilibrium state

The *initial* state $x(t_0) = x_e$ is called a *zero-input equilibrium state* if this state does not change while no input is applied. Formally x_e is defined by

24 $$x(t,t_0,x_e,0) = x_e \qquad \forall t \geq t_0.$$

Since this implies $\dot{x}(t) = 0$ for all $t \geq t_0$, from **8** we obtain for a linear system:

$$P(t)x_e = 0.$$

Hence $x_e = 0$ (the zero-state) is always a zero-input equilibrium state of a linear system, and is unique if $P(t)$ is nonsingular for some $t \geq t_0$.

25 The constant-input equilibrium state for time-invariant linear elements

In the case of a time-invariant system we can also define an equilibrium state for a non-zero constant input $u(t) = u$. Putting $t_0 = 0$, $x(0) = x_0$ we should then have

$$x(t,x_e,u) = x_e \qquad \text{for all } t \geq 0.$$

From **8** we obtain

$$Px_e + Ru = 0,$$

so that for a nonsingular P

26 $$x_e = -P^{-1}Ru$$

is the unique constant-input equilibrium (initial) state of a time-invariant linear system.

27 The zero-input steady state

The *zero-input steady state*, if it exists, is a limiting terminal state when 0 input is applied and if this state x_s is the same for all initial states x_0. Formally

28
$$x_s := \lim_{t \to \infty} x(t, t_0, x_0, 0),$$

if the limit on the right-hand side exists and is independent of x_0. An element either fails to have a steady state, or has a unique one by virtue of the uniqueness of the limit. (This concept of steady state is also called ground state, fundamental state or independent steady state in the literature.)

As we can see from **15** a linear system has a zero-input steady state; namely the zero-state $x_s = 0$ if and only if

29
$$\lim_{t \to \infty} \phi(t, t_0) = 0.$$

The existence of the zero-input steady state is closely connected with the concept of stability to be dealt with in the next chapter.

30 The constant-input steady state of time-invariant elements

In the time invariant case we can also define the steady state for constant-input $u(t) = u$, as

31
$$x_s := \lim_{t \to \infty} x(t, x_0, u)$$

for $t_0 = 0$, if the limit exists and is independent of x_0. From **22** we can conclude that the constant-input steady state exists if and only if

32
$$\lim_{t \to \infty} \exp(Pt) = 0,$$

and in this case

33
$$x_s = -P^{-1}Ru$$

is the constant-input steady state for nonsingular P, which, if it exists, coincides with the constant-input equilibrium state x_e in **26**.

I have apparently defined the same state twice, so let me recall that

x_e is a particular initial state, which always exists for a linear element and need not be unique, while x_s is a terminal state, which need not exist, but is unique if it does.

Both concepts can also be extended to nonlinear elements, but in that case their existence, uniqueness and the relation between them is more complicated than in the linear case discussed here.

34 *The control circuit*

We now suspend the discussion of the state equation of the transfer element and turn our attention to the control circuit. We will see that a control circuit can also be conceived as a transfer element. This enables a straightforward extension of the concepts and propositions which we have just introduced.

I will represent the linear control circuit only in the state equation form starting from Figure 3 with a few modifications expressed in the form of the following assumptions.

35 *Assumption: Both the sensor and the manipulator are identity transfer elements.* Besides the properties mentioned in **4.11** (as applied to a state equation of the controlled subsystem) this assumption also implies that all the state variables are observed. This arrangement is often referred to as "state-feedback" as contrasted with "output feedback". (Cf. also **13**)

36 *Assumption: The controller contains a subtracting transfer element,* which receives as input signal the vector of controlled variables and the command (or reference) vector and emits their difference: *the actuating signal* (because this difference actuates the controller). This assumption requires that to each scalar component of the state vector, one and only one component of the command input vector should be assigned.

37 *Assumption: The external effect may affect the controller directly.* This is a purely technical assumption which facilitates the transcription of a system to state equation form. It does not imply that the external effect is observable.

38 *Notation*

$y :=$ the vector of the controlled variables

$y^* :=$ the vector of command signals

$v :=$ the vector of the manipulating variables

$z :=$ the vector of external effects.

39 *State equation of the control circuit*

Under assumptions **35-37** the control circuit can be written in the following state equation form

40 $\qquad \dot{y} = P_1 y + P_2 v + R_1 z \qquad$ (controlled subsystem)

41 $\qquad \dot{v} = P_4 v + P_3(y - y^*) + R_2 z \qquad$ (controller)

Thus if we consider the controlled subsystem and controller separately as a pair of transfer elements in state equation form, then y is the state (= output) vector, v and z are the input vectors of the controlled subsystem, while v is the state (= output) vector of the controller with z and $(y - y^*)$ as input vectors.

42 *The control circuit as a transfer element*

Instead of perceiving the control circuit as a connected pair of transfer elements, we can consider it as a single element.

Let us write **40** and **41** in the following form corresponding to **8**:

$$\begin{bmatrix} \dot{y} \\ \dot{v} \end{bmatrix} = \begin{bmatrix} P_1 & P_2 \\ P_3 & P_4 \end{bmatrix} \begin{bmatrix} y \\ v \end{bmatrix} + \begin{bmatrix} 0 & R_1 \\ R_2 & -P_3 \end{bmatrix} \begin{bmatrix} z \\ y^* \end{bmatrix},$$

where $[y', v']$ is the state vector and $[z', y'']$ is the input vector. This justifies my speaking about the transfer element representation **8** as representing also the control circuit or the control system as a whole.

(In **38-43** we have not specified the dimension of the vectors y, v, or z, nor that of the matrices. It should nonetheless be noted that their dimensions may be different.)

Chapter 7

STABILITY AND VIABILITY

1 *The quality of the control system*

The whole line of reasoning in my treatise is directed towards demonstrating that the same real sphere can be appropriately controlled by controllers of different structure. This implies on the one hand that I am abandoning the idea of claiming with any two controllers that one is better than the other, i.e., I do not assume complete pre-ordering among the controllers as regards their qualities. On the other hand I have to provide criteria by which a given controller can be qualified as "appropriate" or "inappropriate" to the performance of a task. To this end I will introduce the concept of "viability".

Before discussing viability, however, I must deal with another qualification of control systems, stability. The first reason for doing so is that there is a close connection between the two concepts — stability is the most important of the sufficiency conditions of viability, as we will soon see. The second reason is that for the study of stability a well elaborated mathematical theory is available, while viability theory is considerably less developed, and instead of ready-made theorems one must often be satisfied with ad hoc methods. This state of the art explains why the subsequent stability analysis of my models goes deeper than the viability analysis. But this reversal should not obscure the fact that viability is the target and stability the means.

A few mathematical concepts should, however, be introduced prior to the definition of stability.

2 The norm of a vector

The norm of a (real or complex) vector b is a non-negative number denoted by $\|b\|$ having the following properties:

a) $\|b\| > 0$, if $b \neq 0$ and $\|0\| = 0$

b) $\|\beta b\| = |\beta| \cdot \|b\|$ for any real or complex scalar β

c) $\|b_1 + b_2\| \leq \|b_1\| + \|b_2\|$ (triangle inequality).

3 The norm of a square matrix

The norm of a (real or complex) *square* matrix B is a non-negative number denoted by $\|B\|$ having the following properties:

a) $\|B\| > 0$, if $B \neq 0$ and $\|0\| = 0$

b) $\|\beta B\| = |\beta| \cdot \|B\|$

c) $\|B_1 + B_2\| \leq \|B_1\| + \|B_2\|$

d) $\|B_1 B_2\| \leq \|B_1\| \cdot \|B_2\|$.

4 The natural matrix norm induced by a vector norm

Any vector norm $\|b\|$ induces a natural matrix norm $\|B\|$ defined as

5
$$\|B\| := \max_{\|b\|=1} \|Bb\|.$$

Between a vector norm and its natural matrix norm the following inequality holds:

6
$$\|Bb\| \leq \|B\| \cdot \|b\|$$

The most frequently used vector norms are the "cube-norm" the "sphere-norm" and the "octahedron-norm" (these names refer to the shape of the solid figure: $\|b\| \leq 1$ in three dimensional space), and the matrix norms induced by them. To be specific we will use the cube-norm throughout, although in subsequent analysis other norms would serve as well.

7 The cube-norm

The cube-norm (also called ∞-norm) of a k-vector $b := [b^1, b^2,..., b^k]'$, is defined as

$$\text{8} \qquad \|b\|_\infty := \max_j |b^j|,$$

i.e. the maximum in absolute value of the components. $|\cdot|$ denotes absolute value (also called: modulus) of a real or complex number.

The matrix norm induced by the cube-norm for the square matrix $B := [B^{ij}]$ reads:

$$\text{9} \qquad \|B\|_\infty := \max_i \sum_j |B^{ij}|,$$

i.e. the maximum of the row-sums formed from the absolute value of the entries.

In what follows I will simply write $\|\cdot\|$ instead of $\|\cdot\|_\infty$.

10 Remark
The purely mathematical concept of vector and matrix norm has nothing to do with the concept of control by norm (as explained in Chapter 3) when referring to a kind of economic behaviour. This misuse of the word "norm" in a double sense should cause no confusion since the actual meaning will always be clear from the context in which the word appears.

11 Bounded time-functions

The scalar valued, vector valued, or square matrix valued time functions: $\varphi(t)$, $f(t)$, $F(t)$, respectively, are said to be *bounded* on the infinite interval $[t_0, \infty]$ if there is a constant number $\delta < \infty$ such that

$$|\varphi(t)| \leq \delta$$
$$\text{12} \qquad \|f(t)\| \leq \delta \qquad \forall t \in [t_0, \infty].$$
$$\|F(t)\| \leq \delta$$

13 Simple matrices, spectral decomposition

The $k \times k$ matrix B is said to be simple, if it has k *linearly independent eigenvectors*: $\ell^1, \ldots, \ell^k$, satisfying

$$B\ell^i = \lambda_i \ell^i \qquad i = 1, 2, \ldots, k$$

where λ_i ($i = 1, 2, \ldots, k$) is the *eigenvalue* belonging to ℓ^i.

If B is a simple matrix, we can form the nonsingular matrix $L := [\ell^1, \ldots, \ell^k]$ (the *modal matrix*) and the diagonal matrix $\Lambda = \langle \lambda_1, \ldots, \lambda_k \rangle$ (the *spectral matrix*) so that

14 $$B = L\Lambda L^{-1},$$

which is called a *spectral decomposition* of B. The modal matrix and the spectral matrix will generally have complex entries even if B is a real matrix.

In the spectral decomposition the spectral matrix is unique up to a permutation of the eigenvalues. But the modal matrix is non-unique. We shall denote by $\mathcal{L}(B)$ the *set of modal matrices* belonging to B:

15 $$\mathcal{L}(B) := \{L \mid L^{-1}BL = \Lambda, \ \Lambda \text{ diagonal}\}.$$

With this notation we can say that the matrix B is simple if and only if $\mathcal{L}(B)$ is non-empty.

A sufficient (but not necessary) condition of B being simple is that all its eigenvalues are distinct: $\lambda_i \neq \lambda_j$ if $i \neq j$.

Furthermore if **14** is a spectral decomposition of B, then a spectral decomposition of $\exp(Bt)$ is:

16 $$\exp(Bt) = L \exp(\Lambda t) L^{-1},$$

a well-known result which will be used frequently in what follows.

17 On the stability analysis of time-invariant systems

In the sequel we will restrict our attention to time-invariant systems in the state equation form

18 $$\dot{x} = Px + Ru, \qquad x(0) = x_0, \ t \geq 0$$

where x is a k-vector, u is an m-vector function of time.

STABILITY AND VIABILITY

It is to be noted, however, that the definitions below are also applicable to time-variant systems and some theorems can also be extended to them with appropriately modified assumptions.

19 Instability

The system represented by the state equation **18** is said to be *unstable* if for some initial state x_0 and constant input $u(t) = u$ the state vector $x(t,t_0,u)$ is not bounded for $t \to \infty$. (Cf. **6.22**)

20 Instability theorem

The system **18** is unstable if and only if the system matrix P has an eigenvalue with positive real part or a purely imaginary multiple eigenvalue.

21 Stability

The system **18** is said to be *stable* if it has a steady state for any initial state x_0 and any constant input $u(t) = u$. (Cf. **6.30**)

(This property is usually called "global asymptotic stability". Since I do not use any other stability concept this condensed term seems permissible.)

Stability implies **6.32** and conversely.

22 Stability theorem

The system **18** is stable if and only if all the eigenvalues of the system matrix P have negative real part.

23 The borderline case

It is important to notice that there are systems which are neither unstable (in the sense of **19**) nor stable (in the sense of **21**). There is between them a *borderline case* usually denoted by the clumsy expression "stable in the sense of Lyapunov but not asymptotically stable".

This case can be defined as one in which the system has no steady state, but still the state vector is bounded. This occurs if and only if P has at least one simple imaginary eigenvalue and no eigenvalue of positive real part. I will exclude such systems from discussion for reasons to be explained later.

24 *The response to bounded input*

The concept of stability refers to a property of the system under constant input. However, the constant input case is fairly irrelevant for a control system, since it is not hard to control a system connected to a constant environment. The question thus arises as to how a system which is stable under constant input will behave if subject to variable input. If the input is unbounded we cannot expect any behaviour resembling stability. But with a bounded input we have the following significant result.

25 *Bounded input theorem*

If the system **18** is stable (under constant input) then its response to a bounded input is also bounded.

26 *Remark to Theorem 25*

In spite of the importance of the theorem and although it is usually not dealt with in the textbooks, (Cf. however, Exercise 13 on p. 388 of ZADEH - DESOER, 1963), I omitted its formalization and do not provide a proof. My reason for doing so is that in the Viability Theorem I will provide a more forceful result; the validity of **25** for simple P is a consequence thereof.

27 Concluding remarks on stability from the economist's point of view

Let us confront the stability concept as introduced in the preceding control theoretical treatment with "economic stability" as perceived by an economist. I refrain from precisely defining this latter concept and try rather to approach it from the opposite side.

An unstable control system cannot represent a stable economy, since in such a control system at least one of the controlled variables goes to plus or minus infinity, or the amplitude of its oscillation does, even in the case of zero input (whenever the system is removed from the equilibrium state). Such a system evidently qualifies as "unacceptable" for an economist.

If the control system is that which I characterized as the borderline case, it may behave in the same way as soon as an infinitesimal *interior* disturbance occurs in the controller, i.e. if the behaviour of the agents changes a little so as to transgress the "borderline" in the direction of instability. This is why I have not dealt with the borderline case.

Hence a "stable economy"—for any meaningful interpretation—must be represented by a stable control system.

However, stability—in its definition—refers only to behaviour in the distant future and only to the constant input case. The economist will certainly object to describing as stable a system which fluctuates violently in the early periods even if this fluctuation subsides with the passage of time. The economist may be more interested in what happens during the early transient period than in the asymptotic behaviour, and more in variable than in constant input.

This is why I suggest that stability of the control system is a necessary but not sufficient condition for the models being acceptable to the economist, and a further qualification is called for; viability.

28 Motivations for the viability concept

Although for a time-invariant linear system stability is a sufficient condition of bounded response to a bounded input, we cannot rest content with this qualitative result. In respect of most economic systems we must assume that the state trajectory must satisfy some given *constraints*, and that the admissible inputs are restricted.

Such constraints cannot be assumed away if modelling economic systems. Without aiming at completeness I will list a series of constraints whose neglect may jeopardize the worth of the analysis.

(i) Negative values of many economic variables have no meaningful interpretation. Such variables are for instance: production, inventory, manpower, consumption, price. Thus they require that a non-negativity constraint be introduced.

(ii) Production may be constrained not only from below—to being non-negative—but also from above by the existing productive capacity and available resources.

(iii) Actual commodity transfers may be constrained by insufficient demand or supply.

(iv) In a monetized economy all or some of the agents are usually constrained by budget constraints, and cannot spend in excess of the cash plus credit at their disposal.

(v) Policy requirements may be set as tolerance limits to e.g. unemployment, inflation, growth rate, balance of payments, welfare targets.

This list should not be considered as a promise that I will take all these constraints explicitly into account in my models. Rather I use them as pointers to a more abstract treatment.

29 *The state trajectory and the input trajectory*

We are still dealing with the time-invariant system **18** and introduce for sake of brevity the following notation:

$$\tilde{x} := \{x(t) \mid t \in [0, \infty)\}, \text{ the } \textit{state trajectory}$$

30

$$\tilde{u} := \{u(t) \mid t \in [0, \infty)\}, \text{ the } \textit{input trajectory}.$$

If a pair of trajectories $(\tilde{u}, \tilde{x})$ satisfies equation **18** it is called a *solution* and we say that the input trajectory $\tilde{u}$ *generates* the state trajectory $\tilde{x}$.

31 The admissible input set and the viability set

Let us be given a subset $\mathcal{U}$ of the control space: the *admissible input set*, and a *proper* subset $\mathcal{X}$ of the state space: the *viability set*. An input trajectory $\tilde{u}$ is said to be *admissible* if $\tilde{u} \subset \mathcal{U}$, and a state-trajectory $\tilde{x}$ is said to be *viable* if $\tilde{x} \subset \mathcal{X}$. A *solution* $(\tilde{u}, \tilde{x})$ is said to be *viable* if $\tilde{u} \in \mathcal{U}$ and $\tilde{x} \in \mathcal{X}$.

32 The viability problem

With the above definitions (which can without difficulty be extended to more general systems than **18**) we can define the viability problem as follows:

Find a subset $\mathcal{U}^* \subset \mathcal{U}$ such that the solution $(\tilde{u}, \tilde{x})$ of **18** is viable for any $\tilde{u} \subset \mathcal{U}^*$. Or formally:

33
$$\tilde{u} \subset \mathcal{U}^* \Rightarrow \tilde{x} \subset \mathcal{X}.$$

As this definition of the viability problem shows, we only look for *sufficient* (and not necessary) viability conditions. Conditions which are both sufficient and necessary seem to be out of reach at the present stage, if we want to produce not only qualitative results but something calculable. Even so, we are eager to make $\mathcal{U}^*$ potentially large so as to enhance our chance of finding viable solutions.

34 The degree of stability and the modal condition number

In the viability analysis that follows two parameters depending on the system matrix P play an important role.

Let $\{\lambda_1, \ldots, \lambda_k\}$ be the set of eigenvalues of P. The number $\mu(P)$ is called *the degree of stability* of P:

35
$$\mu(P) := \min_j \mathrm{Re}(-\lambda_j),$$

i.e. the smallest among the real parts of the negative of the eigenvalues.

According to theorem **22** the system **18** is stable if and only if

36
$$\mu(P) > 0.$$

If P is simple (see **13**), the set of modal matrices $\mathcal{L}(P)$ is non-empty and the modal matrices are nonsingular by definition. Hence the number $\kappa(P)$ called the *modal condition number* of P is well-defined:

37 $$\kappa(P) = \inf \{ \|L\| \cdot \|L^{-1}\| \mid L \in \mathcal{L}(P) \}$$

For any simple matrix P we have

38 $$1 \leq \kappa(P) < \infty$$

since $\|L\| \cdot \|L^{-1}\| \geq \|LL^{-1}\| = \|E\| = 1$.

It is to be noted that the minimization problem **37** is not easy to solve without the specification of the chosen norm. But for the cube-norm $\|\cdot\|_\infty$ (**9**) and some other norms an explicit solution of **37** is given by

39 $$\kappa(P) = \|L^+ K^+\|,$$

where $L \in \mathcal{L}(P)$ is any modal matrix of P, $K = L^{-1}$ and B^+ is the matrix formed from the moduli of the entries of the matrix B : $B^+ := [\,|B^{ij}|\,]$. This problem has been discussed and **39** proved elsewhere under the assumption that P has distinct eigenvalues. (MARTOS, forthcoming.)

The estimation in the following lemma forms the basis of the subsequent Viability Theorem.

40 *Lemma* If the system **18** is stable (**36**) and the system matrix P is simple (**13**) then

41 $$\int_0^t \|\exp P(t-\tau)\| \, d\tau \leq \frac{\kappa}{\mu} [1 - \exp(-\mu t)] < \frac{\kappa}{\mu}, \quad \forall t \in (0, \infty),$$

where the argument P has been omitted from μ and κ.

42 *Proof of Lemma 40*

Let $\mathcal{L}(P)$ be the set of modal matrices of P and $L \in \mathcal{L}(P)$, $P = L \Lambda L^{-1}$. First we will prove:

43
$$\|\exp(Pt)\| \leq \kappa \exp(-\mu t) < \kappa,$$

$$\|\exp(Pt)\| = \|\exp(L\Lambda L^{-1}t\| = \|L\, \exp(\Lambda t) L^{-1}\| \leq \qquad \text{(by 16)}$$

$$\leq \|L\| \cdot \|\exp(\Lambda t)\| \cdot \|L^{-1}\| \qquad \text{by (3d)}$$

Since this holds for any $L \in \mathcal{L}(P)$ we can take the infimum over $\mathcal{L}(P)$ on the r.h.s. and from **37** we obtain:

$$\|\exp(Pt)\| \leq \kappa \|\exp(\Lambda t)\| = \kappa \max_{j} |\exp(\lambda_j t)| = \qquad \text{(by 9)}$$

$$= \kappa \max_{j} \exp[(\operatorname{Re} \lambda_j) t] \qquad \text{by (35)}$$

$$= \kappa \exp(-\mu t)$$

$$< \kappa \qquad \text{(by 36 and } t > 0)$$

Hence **43** holds. In a similar way and by a simple integration we obtain

$$\int_0^t \|\exp[P(t-\tau)]\|\, d\tau \leq \kappa \int_0^t \exp[-\mu(t-\tau)]\, d\tau$$

$$= \frac{\kappa}{\mu}[1 - \exp(-\mu t)] < \frac{\kappa}{\mu}.$$

Hereby **41** is proved.

44 *Assumption* The viability set $\mathcal{X}$ has a non-empty interior: $\overset{\circ}{\mathcal{X}} \neq \varnothing$.

45 *Open ball* For any vector-norm $\|\cdot\|$ let us denote by $\mathcal{B}(b,\delta)$ and call an open ball with *center* b and *radius* δ the set

46
$$\mathcal{B}(b,\delta) := \{y \mid \|y - b\| < \delta\}.$$

If $\|\cdot\|$ is the cube-norm, the open ball $\mathcal{B}(b,\delta)$ is an open cube with center b and edge length 2δ.

If $x_1 \in \overset{\circ}{\mathcal{X}}$ is an interior point of $\mathcal{X}$, there are positive radii $\delta > 0$ such that the open balls $\mathcal{B}(x_1,\delta)$ belong to $\mathcal{X}$. Let us denote by $\delta^*(x_1)$ the supremum of such radii:

47
$$\delta^*(x_1) = \sup \{\delta \mid \mathcal{B}(x_1,\delta) \subset \mathcal{X}\}.$$

We have
$$0 < \delta^*(x_1) < \infty$$

for all $x_1 \in \overset{\circ}{\mathcal{X}}$ since $\mathcal{X}$ was assumed to be a proper subset of the state space.

From **46** and **47** it follows that for any $x_1 \in \overset{\circ}{\mathcal{X}}$

48
$$\mathcal{B}(x_1,\delta^*(x_1)) \subset \mathcal{X}.$$

Furthermore for any trajectory $\tilde{x}$

49
$$\|x(t)-x_1\| < \delta^*(x_1), \quad \forall t \in [0,\infty) \Leftrightarrow \tilde{x} \subset \mathcal{B}(x_1,\delta^*(x_1)).$$

50 *Compatible sets, compatible states*

It goes without saying that for a pair of arbitrary sets $\mathcal{U}$ and $\mathcal{X}$ the system **18** need not have a viable solution. Hence we must necessarily impose some restrictions on the sets $\mathcal{U}$ and $\mathcal{X}$ (with regard to **18**) so that the existence of a viable solution (at least for a set of initial points) can be guaranteed. This restriction is called: *compatibility*.

$\mathcal{U}$ and $\mathcal{X}$ are said to be *compatible* (with respect to **18**) if there is a pair of points $u_1 \in \mathcal{U}$ and $x_1 \in \overset{\circ}{\mathcal{X}}$ satisfying

51
$$Px_1 + Ru_1 = 0,$$

or in other words, the set $\mathcal{U}_1$ of *compatible inputs* is non-empty:

52
$$\mathcal{U}_1 := \{u_1 \in \mathcal{U} \mid -P^{-1}Ru_1 \in \overset{\circ}{\mathcal{X}}\} \neq \emptyset.$$

Let us call

53
$$\mathcal{X}_1 := \{x_1 \mid x_1 = -P^{-1}Ru_1 \text{ for some } u_1 \in \mathcal{U}_1\}$$

the set of *compatible states*, which is a non-empty subset of $\overset{\circ}{\mathcal{X}}$ if and only if $\mathcal{U}_1$ is non-empty, i.e. $\mathcal{U}$ and $\mathcal{X}$ are compatible.

We do not claim that if $\mathcal{U}$ and $\mathcal{X}$ are incompatible then there is no viable solution of **18**. But the compatibility of $\mathcal{U}$ and $\mathcal{X}$ will be another condition of the viability theorem.

54 *Viability Theorem* (See Figure 4)

Let the following conditions hold:

a) The system **18** is stable ($\mu > 0$, **36**).

b) The system matrix **P** is simple ($\mathcal{L}(P) \neq \emptyset$, **37**).

c) The viability set $\mathcal{X}$ has a non-empty interior $\overset{\circ}{\mathcal{X}}$ (**44**).

d) $\mathcal{U}$ and $\mathcal{X}$ are compatible ($\mathcal{X}_1 \neq 0$, **53**).

e) The initial point x_0 is a compatible state ($x_0 \in \mathcal{X}_1$).

Then

(i) There is a number $\delta > 0$ such that any admissible input trajectory $\tilde{u} \subset \mathcal{U}$ satisfying

55
$$\| Ru(t) + Px_0 \| < \delta, \qquad \forall t \in [0,\infty)$$

generates a viable state trajectory: $\tilde{x} \subset \mathcal{X}$.

(ii) For $\delta^*(x_0)$ as defined in **47**, μ and κ as in **36** and **37**

56
$$\delta = \frac{\mu}{\kappa} \delta^*(x_0)$$

validates **55**.

57 *Interpretation of and remarks on the Viability Theorem* **54**

1. Part (i) of the theorem tells us that conditions a) to e) are sufficient for the *existence* of admissible input trajectories generating viable state trajectories, since e.g. $u(t) = u_1$, $\forall t$ is such a trajectory for any solution $u_1 \in \mathcal{U}$ of $Px_0 + Ru_1 = 0$. Under assumptions d) and e) such a u_1 exists. Moreover it tells us that there is a neighbourhood of such a u_1 so that the input trajectories belonging to its intersection with $\mathcal{U}$ also do the job.
   Part (ii) yields a *calculable* bound from which the neighbourhood mentioned above can be determined.

2. Let us have a closer look at conditions a) to e) one by one:
   a) The *stability* condition seems to be indispensable, since this is

Figure 4

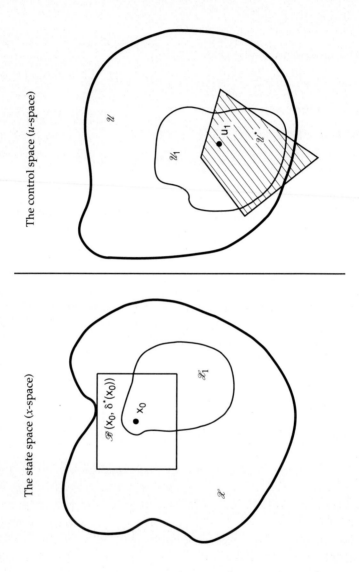

Viability concepts

what guarantees bounded response to any bounded input (Cf. **25**).

b) The *simplicity* of P is a technical matter. I conjecture that this assumption can be discarded without invalidating part (i) of the theorem, while rendering its proof more difficult to obtain. Perhaps an analogue to part (ii) can also be established in this case, but it cannot be so simple.

c) The stipulation that $\mathcal{X}$ should have a *non-empty interior* is essential for the present analysis. Its relaxation to relative interior seems to be attainable.

d) The *compatibility* of $\mathcal{U}$ and $\mathcal{X}$ as defined here can perhaps be relaxed, but some kind of compatibility condition cannot be dispensed with if the existence of viable solutions to **18** is to be guaranteed.

e) The condition $x_0 \in \mathcal{X}_1$ is unnecessarily strong. As I have proved elsewhere it can he relaxed to:

58
$$x_0 \in \overset{\circ}{\mathcal{X}}, \quad \|x_0 - x_1\| < \frac{1}{\kappa} \delta^*(x_1) \quad \text{for some } x_1 \in \mathcal{X}_1,$$

in which case $\delta^*(x_1)$ replaces $\delta^*(x_0)$ in **56**. (MARTOS, forthcoming)

3. The use of the cube norm was chosen for its notoriety. A more flexible kind of norm, whose parameters can be appropriately adjusted, may yield better results. If we know the shape of $\mathcal{X}$, we may want to opt for a norm such that the corresponding open balls would be geometrically more similar to $\mathcal{X}$ than a cube. This idea has been exploited for parallelepiped shaped viability sets in *op. cit.* The choice of the norm is particularly important in connection with Remark 2e), since both **56** and **58** depend very much on it.

4. Parts (i) and (ii) of the theorem have been separated because, when applied to particular economic control systems in this book, part (i) only will be verified (by satisfying conditions a) to e)) while the calculation of δ as under (ii) will not be carried out.

59 Proof of Theorem 54

In the course of the proof we will apply the inequality (KNOBLOCH - KAPPEL, 1974, p. 61):

60
$$\left\| \int_{t_1}^{t_2} f(\tau) d\tau \right\| \leq \int_{t_2}^{t_2} \| f(\tau) \| d\tau, \quad \text{if } t_2 \geq t_1.$$

Let u_1 be a solution of

$$Px_0 + Ru_1 = 0$$

which exists under condition e). In view of 6.21 and 6.22 we have

$$x(t,x_0,u) - x_0 = [x(t,x_0,u) - x(t,x_0,u_1)] + [x(t,x_0,u_1) - x_0] =$$

$$= \int_0^t \exp(P(t-\tau))[Ru(\tau) - Ru_1] d\tau + 0.$$

Hence for all $t > 0$

$$\| x(t) - x_0 \| \leq \left\| \int_0^t \exp[P(t-\tau)] \cdot [Ru(\tau) - Ru_1] d\tau \right\| \leq$$

$$\leq \int_0^t \| \exp[P(t-\tau)] \cdot [Ru(\tau) - Px_0] \| d\tau \leq \qquad \text{(by 60)}$$

$$\leq \int_0^t \| \exp[P(t-\tau)] \| \cdot \| [Ru(\tau) - Px_0] \| d\tau \leq \qquad \text{(by 6)}$$

$$\leq \delta \int_0^t \| \exp[P(t-\tau)] \| d\tau < \qquad \text{(by 55)}$$

$$< \frac{\mu}{\kappa} \delta^*(x_0) \frac{\kappa}{\mu} = \delta^*(x_0). \qquad \text{(by 56 and 42)}$$

From **49** and from $x_0 \in \mathcal{X}_1 \subset \overset{\circ}{\mathcal{X}}$ it follows that $\tilde{x} \subset \mathcal{B}(x_0, \delta^*(x_0))$ and, from **48**, that $\tilde{x} \subset \mathcal{X}$. Hereby we have proved both parts (i) and (ii) of the theorem.

BIBLIOGRAPHICAL NOTES TO PART ONE

The approach to the economy as a control system was pioneered in the fifties by the works of SIMON (1952), TUSTIN (1953), PHILLIPS (1954), GEYER - OPPELT (1957). One of the first — relatively unsuccessful — attempts at a synthesis was provided by LANGE (1965). In the seventies the idea became widespread particularly as a reflection of the developments in the mathematical theory of optimal processes initiated by *Pontryagin, Bellman* and others.

A different start was taken in *Kornai*'s Anti-equilibrium (AE, 1971). Many basic ideas of *Chapter 2* (especially **1-10**) and hence of the whole study can be traced back to this book which I have here followed more or less closely. In the remarks on orderly and orderless markets (**11-15**) I used the works of ARROW - HAHN (1971), BENASSY (1982) and FISHER (1983) among others. The terminology is eclectic, the term "orderly market" and "Hahn process" is used by Fisher, while Benassy uses the expressions "efficient market" and "short-side rule" instead. The discrimination between trading partners (**15c**) as a cause of orderless markets was not to be found anywhere in the disequilibrium literature.

The "control by norm" principle discussed in *Chapter 3* first appeared in the context of economic model construction in SIMON (1952), and was applied in a paper by KORNAI - MARTOS (1973 and NPC, Chapter 2). The principle was thereafter thoroughly discussed by KORNAI (NPC, Chapter 4), my own interpretation differing from his here and there. In sections 3.**5-10** I have relied heavily on KAWASAKI - MCMIILLAN - ZIMMERMANN (1982). To wit, the first (Hungarian) version of this book was already completed when I discovered to my satisfaction how well their findings supported my assumptions. A paper by LACKÓ (1982) also studies the validity of the control by norm principle empirically, but in a macroeconomic setting — allocation of investments among sectors — and hence is more distantly related to my subject.

Chapter 4 is a rephrasing of NPC, Chapter 1, Section 5, by *Kornai* and *Martos*. The concept of vegetative functioning emerged and the term was coined by Kornai in AE and was further refined in the course of our joint work directed to its modelling. (See e.g. NPC, Chapter 2.) The same applies to the five-stage qualification of economic control systems whose first variant appeared in an unpublished manuscript by *Kornai*.

The matrix algebra which is a prerequisite to the understanding of the present book is usually to be found in any matrix theory textbook, e.g. GANTMACHER (1959), LANCASTER (1969). But most textbooks on linear systems also contain an introductory chapter or appendix on matrices.

The basic concepts of systems control in *Chapters 5* and *6* are standard, but the terminology varies from author to author. My main sources were CSÁKY (1977), ELGERD (1967), MELSA - SCHULTZ (1969), ZADEH - DESOER (1963); as concerns anticipatory systems see ROSEN (1985).

These authors, however, define only the zero-input equilibrium state and zero-input steady state since for time-variant systems this is the obvious way to proceed. I, however, had to look ahead to viability analysis and introduce constant-input steady and equilibrium states for time invariant systems.

It is to be seen in *Chapter 7* that this simple extension pays off. While Theorems **20** and **22** are classical results, Theorem **25** and its consequences rely on a means of reasoning which I found only in one passage of Chapter 7 of ZADEH - DESOER (1963). I had to correct their arguments, however, since they use the undefined concept of "norm of a non-square matrix". The bound of Lemma **40** does not appear even there.

The concept of *viability* appeared already in MARTOS - KORNAI (1973 and Chapter 2 of NPC) in a very rudimentary form and was developed further in KORNAI - SIMONOVITS (1977 and NPC, Chapter 10) and SIMONOVITS (NPC, Chapter 13) by the introduction of the local viability concept and the constrained control respectively. The present, more general treatment was very much influenced by the approach taken by AUBIN - CELLINA (1984). A crucial difference should be mentioned, however, in that they work with a closed viability set, and are interested mainly in the change in behaviour when the boundary (between life and death) is hit, while I try to keep the system away from this boundary, and hence closedness need not be assumed.

PART TWO

THE CONTROL OF AN OPEN LEONTIEF - ECONOMY

WITHOUT COORDINATION

This second part of the book presents five variants on how the same real-economy can be operated by fitting to it different controllers. The controllers differ in the input signals and intermediary signals they use in their operation and in the structure of connections between these signals. The emphasis of the analysis is on the operating characteristics (stability and viability) of the control systems. The controllers dealt with here are all vegetative (uncoordinated) ones, while coordinated controllers will be discussed in Part Three.

Chapter 8

THE COMMON REAL SPHERE AND THE COMMON FORM OF THE BEHAVIOURAL EQUATIONS

Real organizations

In Part Two I will study economies where the number of real organizations is n+1, with one exception to which I will shortly return. n is the number of productive firms (or sectors), each producer produces a single homogenous product by a single technology. We can conceive the amount of products to be measured in physical units.

The (n+1)-th real organization, which will occasionally be termed the "consumer", is concerned with "end use" and it is well known from the open Leontief model. The volume of products delivered for end use comprises not only the consumption of the households and public consumption, but all the other uses as well (e.g. for private, public and productive investment, export), except for the material purchases of the productive firms on current account and the change of the product (output) stock of the producers. The consumer is supposed to be autonomous, which means that end use is an exogenously given function of time. (More precisely: it is observable at any point of time, but not controllable.) It belongs to what we called, in the terminology of the control theory, an external effect, load or disturbance coming from the environment. To avoid erroneous connotations of the term "disturbance", I want to stress that we do not postulate any stochastic regularity as regards the end use. In most cases we assume that it is subject to smoothness, non-negativity and boundedness conditions, but to nothing else. This rudimentary treatment is motivated by the fact that the control of interfirm connections is the focus of my attention, and everything else is either represented in a simplified way or neglected completely.

The exception mentioned in the first paragraph is a model where the commercial sector is separated from the productive sectors and handled as a special (n+2)-th real organization. More will be said about this in Chapter 12.

2 Time

Time is represented by the continuous scalar variable t. Thus we will deal with continuously operating controllers. If a variable is set in italics in the formulae, this denotes that it is a function of time, and hence the argument t is usually omitted.

3 *Assumption*:

The time dependent variables are assumed (once and for all) to be continuously differentiable with respect to t in the domain $t > 0$ and differentiable from the right at the initial time $t = 0$. (The time domain $t < 0$ is neglected.) The initial values of the variables at $t = 0$ are considered to be given and denoted by the subscript 0 (e.g. $x(0) = x_0$.)

4 *The end use*

I have already mentioned the only external real process, the *end use*, which is represented by $c :=$ the n-vector of end uses ($c^i :=$ the end use for commodity i.)

It is reasonable to assume that end use is semi-positive, i.e. nonnegative and nonzero : $c \geq 0, c \neq 0$. (With assumption $c = 0$ we would deal with a closed Leontief-model.) Furthermore $c(t)$ is assumed to be bounded in $[0, \infty)$ in the sense of 7.11.

5 *Real activities: production, transfer, stockpiling*

The real activities of the firms are production and transfer of product.

$r :=$ the n-vector of production ($r^i :=$ the amount produced of product i.)

$Y :=$ the n-order square matrix of commodity transfers ($Y^{ij} :=$ the amount of product i transferred from producer i to producer j).

The vector r must be constrained to be non-negative since production is an irreversible process. (Note my words! Production is not *assumed* to be non-negative, it is *stipulated* to be such. We cannot be sure *a priori* that our behavioural equations will not call forth negative production.)

A negative entry of matrix Y can be interpreted as a *return transfer* of goods (to the producer). This is physically possible, thus it need not be totally precluded. If, however, the transfers are decided by the suppliers alone (which is the case in a number of my models), the return of goods cannot be admitted for obvious reasons.

Transfer of goods, or the sum of the transfers of goods to several addressees, is also called "sale" and "purchase" to distinguish as to whether we are looking at this process from the point of view of the supplier or the buyer. But even these terms do not imply (nor, I should add, preclude) the occurrence of a recompensation in exchange for the transfer (e.g. a transfer of money in the opposite direction).

While the variables introduced so far represent flows of commodities, the following represent stocks, the results of the agents' stockpiling activity:

q := the n-vector of output stocks (q^i := the amount of commodity i stored by its producer and available to be transferred).

V := the n-order matrix of input stocks (V^{ij} := the stock of good i stored with producer j waiting for processing in the production of good j).

Both input and output stocks will of course be constrained to non-negative values. Furthermore we have to face the problem of nonstorable commodities (e.g. electricity, services) which would require different handling. But I will disregard these as well as losses occurring in the course of storing perishable goods at this point.

6 *The technology*

The technology applied at time t is represented by the given n-order matrix $A(t)$, which is well-known from the open static Leontief model. Its entry A^{ij} is the amount of product i which is used up in the production of one unit of product j. Thus if r^j is produced of product j, then the amount $A^{ij}r^j$ is used up of material i. In this way the whole matrix of material input can be expressed as the matrix product

$$A\langle r \rangle,$$

where $\langle r \rangle$ is the diagonal matrix formed from the vector r.

7 *Assumptions*: We assume the matrix $A(t)$ to possess the following properties for all $t \geq 0$.

a) *Non-negativity*: $A \geq 0$. This assumption precludes joint products.

b) *Irreducibility*: There is no symmetric permutation of rows and columns of A, such that in the partition:

$$A = \begin{bmatrix} A_{11} & A_{12} \\ A_{21} & A_{22} \end{bmatrix}$$

A_{11} is square, $A_{21} = 0$, A_{22} is non-empty.

This assumption precludes the existence of a group of producers who could do without inputs coming from outside the group.

From assumptions a) and b) it follows that A has a positive dominant eigenvalue and a corresponding positive eigenvector.

c) *Productivity*: The spectral radius ρ of matrix A (which under assumptions a) and b) equals the dominant eigenvalue) is less than one: $\rho < 1$. This assumption implies the productivity of GALE (1980): $\exists x > 0$ such that $x > Ax$.

Under assumptions a), b) and c) the matrix $(E - A)$ is non-singular and its inverse C (the so-called Leontief-inverse of A) is positive:

$$C := (E - A)^{-1} > 0.$$

d) *Distinct eigenvalues*: The matrix A has distinct eigenvalues. This implies that it is simple (Cf. 7.14).

This is a purely technical assumption for the sake of mathematical convenience without economic significance and interpretation (since by a small perturbation any matrix can be made simple).

As a consequence of this assumption the spectral decomposition of A can be given in the form

9
$$A = F\phi F^{-1},$$

which will be frequently made use of. Let me recall that in 9 ϕ is the (diagonal) spectral matrix and F a (non-unique) modal matrix of A.

e) *Time-invariance*: The matrix A is constant in time

$$A(t) = \mathbf{A}, \ \forall t.$$

While the previous assumptions were either "natural" or neutral from an economic point of view, this last one strikes hard at the scope of validity of my models. Although all the models which follow could have been stated without this assumption, and certain of the theorems (e.g. the results of the structural analysis) would have also remained valid (at the expense of some small refinements) for changing technology, and with certain others the assumption could be relaxed. Nonetheless I have maintained this facile assumption for the following reason. It would be hard to follow the main line of thought if for different models — and for their analysis from different aspects — I had to switch from milder to more restrictive assumptions and back. On the other hand I do not want to camouflage the fact that there is a series of economic phenomena (one of them is technological change or, even more so, technological choice) which I was not yet able to handle adequately in the present theoretical framework.

10 Equations of the real processes

The real processes of the system are described by $n+n^2$ scalar equations condensed into an n-vector equation and an n by n matrix equation. The first one portrays the formation and change of output stocks, the second that of input stocks.

THE OUTPUT BALANCE

11
$$\dot{q} = r - Ye - c,$$

i.e. the rate of output stock changes equals: the amounts produced minus the sum of transfers for productive and for end use ($e' := [1, 1,..., 1]$, the summation vector.)

THE INPUT BALANCE

12 $$\dot{V} = Y - A\langle r \rangle$$

i.e. the rate of input stocks changes equals: the amount of input material received minus that used up in the production of r.

Such a representation of the real processes implies the unrealistic assumption that neither the production nor the transfer takes time.

13 *The common form of the controllers*

Having finished the introduction of the equations representing the real processes, which will form the essential part of the controlled subsystem in the models that follow, we will now turn to some common characteristics of the five—in many other aspects different—controllers.

14 *The controllers are linear*

Apparently a strong stipulation, this is not so from the point of view of the analysis I am going to do. Namely if I am able to find linear controllers which work adequately, I will have proved the existence of an appropriate controller in general. It is another question whether there is a nonlinear controller that would perform better from some other aspect, or whether some other models, which could not be appropriately operated by a linear controller might be put aright by the use of a nonlinear one.

15 *The controllers are of the proportional plus integrating (PI) type*
(Cf. 6.4)

This type turned out to be the *simplest* among the linear controllers by which satisfactory operation can be achieved. In the absence of the integrating term viability cannot be achieved without unpleasant additional assumptions and the proportional term is indispensable — as we will soon see — for (asymptotic) stability.

16 *The actuating signal* (Cf. **6.36**)

The actuating signal is the difference between the controlled variables and their (constant) norms. This is one of the mathematical forms, not the only one, in which the control by norm principle (**3.2**) can be expressed. Under this assumption the input signal of the controller (the actuating signal) takes the form:

17
$$y(t) - y^*,$$

where the notation corresponds to that of **6.38**, except that the (time dependent) command signal $y^*(t)$ has been replaced by the constant norm y^*.

Since I refrain from constructing higher order systems (Cf. **5.10**), the only alternative to using constant norms as command signals would have been to define them as exogenously given (e.g. linear or exponential) functions of time. I feel no remorse for the rejection of this alternative, which would have made the analysis (slightly) more difficult without making the model more realistic.

18 *A differential operator*

It follows from assumptions **14** and **15** that the *derivative* of the manipulating variables will be expressed by applying the following differential operator to the actuating signal:

$$\alpha_1 \frac{d(\cdot)}{dt} + \alpha_2 \cdot (\cdot),$$

where the first term represents the proportional (P), the second the integrating (I) action. Since this operator will be provided with positive or negative sign as economic common sense demands, there is no loss of generality in assuming that α_2 is positive. Hence we introduce the notation

$$\alpha_2 =: \gamma^2$$

To facilitate later calculations we introduce also the notation

$$\beta := \frac{\alpha_1}{2\gamma}.$$

In this way we obtain the differential operator

19
$$\Gamma(\cdot) := 2\beta\gamma \frac{d(\cdot)}{dt} + \gamma^2 \cdot (\cdot)$$

to be applied in the controllers. Since the sign of the second term does not depend on the sign of γ, and that of the first can be changed by giving a sign to β, there is again no loss of generality if we assume once and for all that

20
$$\gamma > 0.$$

Finally, if we apply the differential operator **19** to the actuating signal **17**, we obtain the formula:

21
$$\Gamma(y(t) - y^*) = 2\beta\gamma\dot{y} + \gamma^2(y - y^*),$$

which will be used frequently.

It may be noted that the control parameters β and γ are well known from linear control theory, where they are usually called *damping exponent* and *natural frequency*, respectively, in view of the roles they play in the solution of a one dimensional control system.

22 Uniform parameters

It remains an open question whether, in the course of generating different manipulating variables (producing different products, providing different materials to different producers) by the operator Γ —even if the formula of the operator, the behavioural rule is the same—the numerical value of the parameters β and γ must be identical or different. Of course, we have no reason to assume that all the economic agents in all their kinds of economic activity obey behavioural rules "tuned in" by uniform values of the control parameters.

In spite of, and without challenging, this argument I use a uniform value of these parameters by force of the following reasoning. The dialectics are similar to those which I used when arguing in favour of linearity. If we successfully tune the controller when using only two (or elsewhere three) parameters, then a larger number of them may improve its operation without adding anything to the result, i.e. that the system can be appropriately tuned in. If I do not require more from a radio set than to receive the selected station at an adequate volume, then I may be content with two knobs, the tuning knob and the volume control. A third one (e.g. tone controller) may improve the quality of the reception, but its existence is immaterial from the point of view of the first two criteria. This is why we can make do with two parameters (or with three of them, where non-negativity of the prices is an additional criterion) without clashing with the sensible perception that in reality the diverse activities of different economic agents are characterized by many more parameter values.

23 *The reduced form*

The rest of the present chapter is devoted to the study of a specific form of control system. Its name *reduced form* stems from a connection with the models that follow, which (with one exception) will be reduced so as to fit into this form.

Let us consider the following control system, not yet in state equation form, (still we make use of the notation of 6.38).

24 $$\dot{y} = Qv + z \quad \text{(The controlled subsystem)}$$

25 $$\dot{v} = -2\beta\gamma\dot{y} - \gamma^2(y - y^*) \quad \text{(The controller)}$$

where β and γ are familiar from **18**, and **25** is a PI controller.

By substituting $\dot{y}$ from **24** into **25** we get the system in state equation form which corresponds to 6.40-41:

26 $$\dot{y} = Qv + z$$

27 $$\dot{v} = -2\beta\gamma Qv - 2\beta\gamma z - \gamma^2(y - y^*),$$

or equivalently in the form corresponding to 6.43:

28 $$\begin{bmatrix} \dot{y} \\ \dot{v} \end{bmatrix} = \begin{bmatrix} 0 & Q \\ -\gamma^2 E & -2\beta\gamma Q \end{bmatrix} \begin{bmatrix} y \\ v \end{bmatrix} + \begin{bmatrix} E & 0 \\ -2\beta\gamma E & \gamma^2 E \end{bmatrix} \begin{bmatrix} z \\ y^* \end{bmatrix}.$$

Comparing **28** with **6.43** we can identify the matrices and notice two important properties:

(i) In **28** all the matrices are constant, thus we deal with a time-invariant system,

(ii) y and v have the same number of components (k, say), thus all the submatrices in **28** are square (k × k) matrices.

29 *Notation*

a) Denote the eigenvalues of Q by $\sigma_1,\ldots,\sigma_k$ and define the set of eigenvalues (called: the *spectrum* of Q) by:

$$\mathscr{Q} := \{\sigma_1,\ldots,\sigma_k\}.$$

b) Let us define the non-negative real valued function ϑ of the complex variable ζ for Re $\zeta > 0$ as

30
$$\vartheta(\zeta) := \frac{|\operatorname{Im} \zeta|}{2|\zeta|\sqrt{\operatorname{Re} \zeta}}.$$

ϑ maps the open right halfplane of the complex number plane onto the interval $[0, \infty)$ of the real line. Observe that $\vartheta = 0$ for real ζ, hence in the following theorem only the complex eigenvalues of Q constrain β.

31 *Stability theorem of the reduced form*

Suppose Q is a simple matrix. The reduced form **24-25** is stable if and only if

a) Re $\sigma > 0$, $\forall \sigma \in \mathscr{Q}$ and

b) β > max $\{\vartheta(\sigma) \mid \sigma \in \mathscr{Q}\}$.

32 *Comments* (i) The maximum on the right hand side of b) exists since $\mathscr{Q}$ is a finite set and $\vartheta(\sigma)$ is finite for all σ satisfying a).

(ii) The theorem tells us that the stability of the reduced form does not depend on γ. This is achieved by the queer formulation of the controller **25** in accord with the notation of **18**.

(iii) The theorem also tells us that if Q satisfies a) then the system can always be stabilized by choosing β sufficiently large, the lower bound for β depending only (and rather complicatedly) on the spectrum of Q.

33 *Proof of theorem* **31**

In view of the stability theorem **22** we have to establish the necessary and sufficient conditions for the eigenvalues of the system matrix of **28**

$$P := \begin{bmatrix} 0 & Q \\ -\gamma^2 E & -2\beta\gamma Q \end{bmatrix}$$

to have negative real part. In the proof we will rely on well-known formulae concerning determinants, especially those of hypermatrices with commuting blocks:

$$\begin{vmatrix} A & B \\ C & D \end{vmatrix} = |AD - BC|.$$

The eigenvalues $\omega_1,...,\omega_{2k}$ of P are the solutions of the (2k)-order characteristic equation:

$$|\omega E - P| = 0.$$

Let $Q = S\Sigma S^{-1}$, $\Sigma := <\sigma_1,...,\sigma_k>$ be a spectral decomposition of Q.
In partitioned form we have

$$\begin{vmatrix} \omega E & -Q \\ \gamma^2 E & \omega E + 2\beta\gamma Q \end{vmatrix} = |\omega^2 E + 2\beta\gamma\omega Q + \gamma^2 Q| =$$

$$= |S^{-1}| \cdot |\omega E + 2\beta\gamma\omega Q + \gamma^2 Q| \cdot |S| =$$
$$= |\omega^2 E + 2\beta\gamma\omega\Sigma + \gamma^2\Sigma| = 0.$$

This (2k)-degree equation in ω reduces to quadratic equations of number k:

$$\omega^2 + 2\beta\gamma\sigma\omega + \gamma^2\sigma = 0 \quad \sigma \in \mathscr{Q}$$

yielding a solution pair

34
$$\omega_{1,2} = \gamma(-\beta\sigma \pm \sqrt{\beta^2\sigma^2 - \sigma}\,)$$

for each $\sigma \in \mathscr{Q}$.

With the following notation:

$$\mu = \text{Re}\,\sigma$$
35
$$\upsilon = \text{Im}\,\sigma$$
$$\zeta = \beta^2\sigma^2 - \sigma.$$

and by $\gamma > 0$, the condition $\text{Re}\,\omega < 0$ can be expressed from 35 as:

36
$$\pm\,\text{Re}\,\sqrt{\zeta} < \beta\mu,$$

which cannot hold for both signs unless:

37
$$\beta\mu > 0.$$

Let us apply the formula of *Moivre* to obtain

38
$$\text{Re}\,\sqrt{\zeta} = \sqrt{\tfrac{1}{2}(|\zeta| + \text{Re}\,\zeta)}\,,$$

put this into 36, take the square of both sides and rearrange:

$$|\zeta| < 2\beta^2\mu^2 - \text{Re}\,\zeta.$$

The right hand side must be positive, but this will be guaranteed by 40, so we need not care about it here, and take the square once more to obtain:

$$(\text{Re}\,\zeta)^2 + (\text{Im}\,\zeta)^2 < 4\beta^4\mu^4 - 4\beta^2\mu^2 \text{Re}\,\zeta + (\text{Re}\,\zeta)^2.$$

We substitute here according to 35

$$\text{Re}\,\zeta = \beta^2(\mu^2 - \upsilon^2) - \mu$$
$$\text{Im}\,\zeta = \upsilon(2\beta^2\mu - 1)$$

which after rearrangement gives

39 $$\upsilon^2 < 4\beta^2\mu(\mu^2 + \upsilon^2).$$

This cannot hold unless

40 $$\mu = \operatorname{Re} \sigma > 0$$

by which we proved the necessity of **31a)**, and furthermore with **37** we have

41 $$\beta > 0$$

as another necessary stability condition. Thus from **39** and **40** we get

$$\beta^2 > \frac{\upsilon^2}{4(\mu^2 + \upsilon^2)\mu} = \frac{(\operatorname{Im} \sigma)^2}{4|\sigma|^2 \operatorname{Re} \sigma}$$

which must hold for each $\sigma \in \mathcal{Q}$. We take the square root and in view of **41** and **30** conclude that **31b)** is, indeed, a necessary condition. The way we deduced necessity proves that the two conditions **31a)** and **b)** taken together are also sufficient for the stability of the reduced form.

* * *

Having finished with the discussion of some common features of the five models to follow, I will now turn to their individual, distinctive characteristics.

Chapter 9

THE STOCK SIGNAL
(MODEL S)

1 Inventories

The holding of stocks of goods (inventories) is a thoroughly explored economic activity. The literature dealing with inventory research is plentiful, and there even exists an *International Society for Inventory Research*. In the literature we find many kinds of classifications of stocks, e.g. according to where they are (stocks of input materials and intermediate products, intermanufacturing stocks of semi-finished products, stocks of finished products or output stocks, wholesale and retail commercial stocks, stocks in transportation, stocks at the final user), for what purpose they are held (working stocks, buffer stocks for peak loads, replenishment period stocks, batch size stocks if order sizes are fixed, speculative stocks, strategic stocks) and others. The most important part of the research has been devoted to the explanation of how and why the inventories change on the macro or micro level, of the interrelation between inventory policy and business cycles and — from a normative aspect — of how the optimum size of inventories can be determined by mathematical methods of operations research.

2 Stock signals

A far less frequently discussed topic is the role played by the information on inventories (called: stock signals in this book) in the control of the economic processes. I am going to show the importance of this signal by demonstrating in this chapter that many basic economic processes (like production level and commodity transfer among the producers) can be controlled by a vegetative controller relying *exclusively* on stock signals, and that such a system is viable.

Among the many kinds of stocks I deal here only with those which have appeared already in equations 8.11 and 8.12: the output stocks (q) and the input stocks (V). In the controller the information about these controlled variables, which are measured error-free, (see Definition 4.11 and Assumption 6.35) is fed back into the control of production and transfer; the manipulating signals.

3 The behavioural equations of Model S

Making use of the differential operator Γ as defined in 8.19, and using as actuating signal the difference between the stock variables (q, V) and their norms ($\mathbf{q}^*$, $\mathbf{V}^*$) as explained in 8.16, we obtain the following behavioural equations.

PRODUCTION CONTROL

4
$$\dot{r} = -2\beta\gamma\dot{q} - \gamma^2(q - \mathbf{q}^*)$$

CONTROL OF TRANSFERS

5
$$\dot{Y} = -2\beta\gamma\dot{V} - \gamma^2(V - \mathbf{V}^*).$$

These equations form the controller, which we have fitted to the controlled subsystem represented by equations 8.11 and 8.12 to obtain the system called Model S (S for stock). The system is illustrated also by the block diagram in Figure 5, which shows a single-loop system.

6 Interpretation of the behavioural equations

The economic control system represented by Model S can be interpreted the following way:

a) The producers keep output stocks of their own products and input stocks of the materials they use. These stocks will be required to be non-negative as a matter of course.

b) The producers as sellers try to keep their output stocks at a normal level so as to be able to satisfy the buyers. Consequently all the output stock norms are assumed to be positive: $\mathbf{q}^* > 0$.

Figure 5

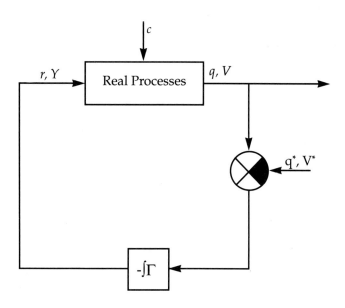

Block diagram of Model S

c) The producer as consumer of input materials tries to keep input stocks of the materials *he himself uses in production* at a normal level so as to satisfy the needs of the production process. The input stock norms of these materials are assumed to be positive: $A^{ij} > 0$ implies $V^{*ij} > 0$. (For $A^{ij} = 0$ see Assumption 7 below.)

d) The level of production is decided by the producers who take into account their output stock norms, the actual value of these stocks and the rate of their change. If the actual output stock exceeds the normal level they will decrease production (and *vice versa*). Furthermore, increasing this stock will exert a similar effect (in the case of $\beta > 0$, which we have not postulated). The negative signs on the r.h.s. of equation 4 correspond to this logic.

e) The decision on transfers is made by the *buyers* relying on principles in parallel to those under d), but, of course, on the basis of the observed and the normal values, as well as the rate of the input stocks.

f) Strictly speaking it cannot be verified that in their decisions on production and purchases producers behave as has been described above. But such behaviour does not conflict with common sense; on the contrary, it is logically plausible. It is also empirically supported by the findings discussed in 3.9. Recall the observed dependence of production on output stocks. The particular mathematical form of the controller, however, has been chosen with regard to analytical convenience.

g) We may also ask the question: what kind of market is represented by the model as a whole? In this market all the purchasing decisions of the buyers can be fulfilled, since there is enough output stock to satisfy buyers' demand. In other words the "short-side rule" is valid in this economy in the sense that it is always the buyer who is on the short side. Thus Model S represents an orderly market, and especially a buyer's market for each commodity.

7 *Assumptions and notation concerning irrelevant variables*

If producer j does not use material i ($A^{ij} = 0$) then we assume that $V^{*ij} = 0$, i.e. the corresponding input stock norm is zero as well. Furthermore it is self-evident to assume that producer j neither has initial stock, nor makes initial purchase of this material ($V_0^{ij} = 0$, $Y_0^{ij} = 0$). Under these assumptions the corresponding components of equation 8.12 and equation 5 reduce to the system

$$\dot{V}^{ij} = Y^{ij}$$

$$\dot{Y}^{ij} = -2\beta\gamma \dot{V}^{ij} - \gamma^2 V^{ij}$$

with the unique solution:

$$V^{ij}(t) = 0, \qquad Y^{ij}(t) = 0 \qquad \forall t \geq 0.$$

Thus the producer will never buy this material and will never have a stock of it. Variables which are thus kept permanently at 0 level are said to be *irrelevant*, and the rest *relevant*.

The equations determining irrelevant variables could simply be crossed out from the system without changing the results of the analysis, so long as A remained irreducible. Still, the explicit distinction between relevant and irrelevant variables would imply inconvenient notation and clumsy formulation in what follows. Therefore I will take the liberty of not paying great attention to the above "refinement", and of handling relevant and irrelevant variables alike.

But occasionally — as a warning — I will use the superscript $^{r\ell v}$ to indicate that a relation is valid for the relevant variables only e.g.

8
$$V^* > {}^{r\ell v} 0,$$

or verbally "V^* is $r\ell v$-positive", will be used as a shorthand for the implication:

$$\begin{cases} A^{ij} > 0 \Rightarrow V^{*ij} > 0 \\ A^{ij} = 0 \Rightarrow V^{*ij} = 0. \end{cases}$$

After this notational digression let us now return to the discussion of the model.

9 Structural analysis

In order to analyse the information and decision structure of the controller let us write equations 4 and 5 in scalar form:

10
$$\dot{r}^i = -2\beta\gamma \dot{q}^i - \gamma^2(q^i - q^{*i})$$

11
$$\dot{Y}^{ij} = -2\beta\gamma \dot{V}^{ij} - \gamma^2(V^{ij} - V^{*ij}) \qquad i,j = 1,2,\ldots,n$$

Equation 10 tells how producer i decides on the production level. For this purpose he has to observe his actual output stock q^i and its rate of change $\dot{q}^i$. These data are available to him by *introspective* observation (Cf. Assumption 4.9). It has been assumed that he also knows his output stock norm q^{*i}. Thus he does not receive any signal from an outside organization, neither does he transmit any to others. The information flow is therefore *non-communicative*. Finally he decides this way on the change of his production level $\dot{r}^i$ without any coordination; the decision process is *uncoordinated*.

The situation is similar as regards equation 11, where producer j decides in a *non-communicative* and *uncoordinated* way on the purchase of material i, relying on the *introspective* observation of his actual input stock V^{ij} and its rate $\dot{V}^{ij}$.

Summing up: the controller of Model S realizes *vegetative non-communicative* control, the "first stage" (4.13). Even more can be said: the purchasing decisions on different materials are separated both from each other and from the decision on production. In this way there is no need to coordinate the activities within a firm between the purchasing agents responsible for the acquisition of different materials and the production manager, nor to establish communication between them.

2 Stability analysis

We will analyse the stability of Model S in two steps. First we will transform the system to the reduced form 8.24-25 by aggregation and establish its stability on the basis of Theorem 8.31. From this the stability of the original system concerning variables q and r follows directly, while stability in the variables V and Y will be analysed in the second step.

13 Transformation to the reduced form

Let us rewrite the equations of the controlled subsystem 8.11-12, the latter being multiplied from the right by the summation vector e (aggregation across users of the same material):

14 $$\dot{q} = r - Ye - c$$

15 $$\dot{Ve} = Ye - Ar.$$

Introducing the notation:

16 $$y = \begin{bmatrix} q \\ Ve \end{bmatrix}, \quad v = \begin{bmatrix} r \\ Ye \end{bmatrix}, \quad z = \begin{bmatrix} -c \\ 0 \end{bmatrix}, \quad y^* = \begin{bmatrix} q^* \\ V^*e \end{bmatrix}$$

17 $$Q^S = \begin{bmatrix} E & -E \\ -A & E \end{bmatrix}$$

we obtain **14-15** in the form

18 $$\dot{y} = Q^S v + z,$$

which corresponds to 8.24, the superscript S referring to Model S.

Applying the notation **16** to the behavioural equations **4** and **5**, after aggregating the latter via a post-multiplication by e, we obtain:

19 $$\dot{v} = -2\beta\gamma\dot{y} - \gamma^2(y - y^*),$$

which corresponds to **8.25**.

20 Stability analysis of the reduced form

In view of theorem 8.31, in order to establish the stability of the system **18-19** we will prove that:

a) If A is nonsingular, then Q^S is simple

b) Re $\sigma > 0$, $\forall \sigma \in \mathscr{2}^S$.

Then, by choosing β according to the inequality 8.31b), the stability conditions will be satisfied.

Let us start with proposition b). The characteristic equation of Q^S can be expanded by the use of the spectral decomposition $A = F\phi F^{-1}$ (Cf. 8.9) in the following way:

$$0 = |\sigma E - Q^S| = \begin{vmatrix} (\sigma-1)E & E \\ A & (\sigma-1)E \end{vmatrix} =$$

$$= |(\sigma-1)^2 E - A| = |F^{-1}| \cdot |(\sigma-1)^2 E - A| \cdot |F| =$$

$$= |(\sigma-1)^2 E - \phi|.$$

This $(2n)$-degree equation reduces to n quadratic equations in σ:

21 $$(\sigma-1)^2 = \varphi \qquad \varphi \in \mathcal{F}$$

where $\mathcal{F}$ is the spectrum of A.

From **21** we obtain

$$\text{Re } \sigma = \text{Re } (1 \pm \varphi^{\frac{1}{2}}) = 1 \pm \text{Re } \varphi^{\frac{1}{2}} \geq 1 - |\varphi^{\frac{1}{2}}|$$

$$\geq 1 - \rho^{\frac{1}{2}} > 0,$$

since $\rho < 1$ by Assumption 8.7c). Hereby proposition b) is proved. Let us now turn to proposition a).

a) Since A is nonsingular, $0 \notin \mathcal{F}$ and **21** has two different roots for any $\varphi \in \mathcal{F}$. Since A has n distinct eigenvalues by Assumption 8.7d) and **21** excludes equal σ's for unequal φ's, we conclude that Q^S has $2n$ distinct eigenvalues, and hence is simple.

22 *Stability analysis of the original system*

From the stability of the reduced form in y and v it follows that the original system is stable in the variables $q = [E\ 0]y$ and $r = [E\ 0]v$, the output stocks and production levels. However, as regards the input stocks V and transfers Y, we can only infer the stability of their aggregates (across users) which we cannot be satisfied with. So let us pick out a pair of scalar equations from matrix equations 8.12 and 5:

$$\dot{V}^{ij} = Y^{ij} - A^{ij} r^i$$

$$\dot{Y}^{ij} = -2\beta\gamma \dot{V}^{ij} - \gamma^2(V^{ij} - V^{*ij}).$$

This pair of equations (in V^{ij} and Y^{ij}) can be considered as a degenerate reduced form with $Q = [1]$, the first order unit matrix. Its single eigenvalue is $\sigma = 1$, so that 8.31 b) reduces to $\beta > 0$, which must be satisfied anyway.

We have thus proved the following theorem.

23 *Stability theorem of Model S*

Consider Model S consisting of the controlled subsystem 8.11-12 and the controller 4-5. Assume that A is nonsingular and Assumptions 7 hold. The system is stable if and only if β satisfies the inequality

$$\beta > \max\{\vartheta(\sigma) \mid \sigma \in \mathscr{L}^S\},$$

where $\mathscr{L}^S$ is the spectrum of Q^S as defined in **17** and $\vartheta(\cdot)$ is defined in **8.30**.

24 *An example*

We will calculate the lower bound for β, taking A from the 12 sector input coefficient matrix of Japan (TSUKUI, 1968). Ten of its eigenvalues are real and positive, they yield real σ and $\vartheta(\sigma) = 0$. From the remaining complex pair of eigenvalues φ we calculate the corresponding four σ's (**21**) and evaluate $\vartheta(\sigma)$.

$$\varphi = 0.158 \pm 0.053\iota \qquad |\varphi| = 0.167$$

$$\sigma_1 = 1.403 \pm 0.067\iota \qquad |\sigma_1| = 1.405 \qquad \vartheta(\sigma_1) = 0.020$$

$$\sigma_2 = 0.597 \pm 0.067\iota \qquad |\sigma_2| = 0.601 \qquad \vartheta(\sigma_2) = 0.072$$

If we were to stabilize Model S for the Japanese A we would have to choose β satisfying:

$$\beta > 0.072 \ .$$

(In engineering practice β is usually chosen from the interval $[0.5, 2]$).

25 *The constant-input steady state*

For a stable system it is interesting to know its constant-input steady state. (See 6.30) Since for the model under discussion we have already assumed that a part of the input signals, the stock norms, are constant, we now have to add that we calculate the system's steady state subject to constant end use:

26 $$c(t) = c \qquad \forall t \geq 0.$$

Since the constant-input steady state (if it exists) coincides with the equilibrium state, (see 6.**24**) it is characterized by

27 $$\dot{q} = 0, \quad \dot{V} = 0, \quad \dot{r} = 0, \quad \dot{Y} = 0 \qquad \forall t \geq 0.$$

Substituting into **4** and **5** from **27**, we get the equalities:

28 $$q_e = q^*, \quad V_e = V^*,$$

i.e. the steady-state (= equilibrium, marked by subscript e) of the stocks coincides with their norm. This is an important criterion, often expressed as the absence in the system of "control-error".

Substituting again from **27** into 8.**11-12** we obtain the algebraic equation system

$$r_e - Y_e e = c$$
$$Y_e - A\langle r_e \rangle = 0,$$

whose solution is:

29
$$r_e = Cc$$
$$Y_e = A\langle Cc \rangle$$

where $C := (E - A)^{-1}$, the Leontief-inverse.

These two components of the equilibrium state can be interpreted as r_e being the "total production", satisfying the constant end use c, and Y_e the matrix of material inputs corresponding to this production level.

This equilibrium state is non-Walrasian since positive output stocks and input stocks persist which are not needed to sustain the equilibrium production and turnover of the commodity bundle required for end use.

30 *Viability analysis*

Having been successful in establishing the stability of Model S, I wish to remind the reader that we cannot rest content with this result. The real problem is that of establishing viability, of which stability is only the first — albeit fundamental — condition.

The analysis will be carried out on the basis of the Viability Theorem 7.54. We have only to find out under what specifications of Model S conditions a) to e) of the Viability Theorem will be satisfied.

a) The *stability* conditions of Model S have been settled in Theorem **23**.

b) The *simplicity* of the system matrix Q^S of the *reduced form* was established in **20** a) under the assumption that A is nonsingular. Although we do not know whether or how this extends to the system matrix of Model S — so large that I did not write it out in full — in view of Remark 7.57 2 b) I take the liberty of not checking this property. It would in any case be an enormous task yielding no economic lesson.

In view of this remark I *shall disregard the simplicity assumption b) in the analysis of the subsequent models as well*. However, the simplicity of the system matrix of the reduced form must in any case be established for the stability analysis.

Consequently, the following tasks are still before us:

c) To define the viability set $\mathcal{X}^S$ (with non-empty interior) and the admissible input set $\mathcal{U}^S$.

d) To show that $\mathcal{U}^S$ and $\mathcal{X}^S$ are compatible and to determine the set $\mathcal{U}^S_1, \mathcal{X}^S_1$ of compatible inputs and states, respectively.

e) To derive the conditions under which the initial state is compatible.

Our state variables are q, V, r and Y, the inputs q^*, V^* and c of which q^* and V^* are constant. Since in this model the transfer is decided by the buyer, *return transfer* (8.5) will be *admitted* and Y can take any value. Hence it will be omitted in the definition of the viability set.

31 *Definition of the viability set and the admissible input set*

The *viability* set will be defined according to the following restrictions

i) The input and output stocks should be non-negative.

ii) The production should be non-negative and cannot exceed the given productive capacity $\bar{r} > 0$.

32 $$\mathcal{X}^S := \{q, V, r \mid q \geq 0, V \geq 0, 0 \leq r \leq \bar{r}\}.$$

Since $\bar{r} > 0$ the viability set has a non-empty interior, as stipulated by **7.54** c). The interior of $\mathcal{X}^S$ is:

33 $$\overset{\circ}{\mathcal{X}}{}^S := \{q, V, r \mid q > 0, V >^{r\ell v} 0, 0 < r < \bar{r}\}.$$

The upper bound $\bar{r}$ in **32** expresses a constant capacity restriction on the level of production. This is an indirect representation of the fact that any production process draws on primary resources which are scarce. The facile assumption that $\bar{r}$ is constant is not really necessary and it can be replaced by an exogenously given time function $\bar{r}(t)$ without much difficulty. But the real problem of endogenising $\bar{r}(t)$ by converting a part of the end use to net investment is beyond the scope of the present study.

The admissible input set will be so defined that:

iii) The output stock norms will be positive and input stock norms $r\ell v$-positive.

iv) The end use will be semi-positive.

34 $$\mathcal{U}^S := \{q^*, V^*, c \mid q^* > 0, V^* >^{r\ell v} 0, c \geq 0, c \neq 0\}.$$

35 *Compatibility*

We claim that a point of $\mathcal{U}^S$ is a compatible input if and only if

36 $$Cc < \bar{r},$$

i.e. the set of compatible inputs is:

37 $$\mathcal{U}_1^S := \{q_1^*, V_1^*, c_1 \mid q_1^* > 0, V_1^* >^{r\ell v} 0, c_1 \geq 0, Cc_1 < \bar{r}, c_1 \neq 0\}.$$

The set $\mathcal{U}_1^S$ is non-empty since $\bar{r} > 0$ implies that **36** can be satisfied by some sufficiently small semi-positive c_1. To show that $\mathcal{U}_1^S$ is indeed the set of compatible inputs we should recall that the states to be checked are the steady states corresponding to constant inputs from $\mathcal{U}_1^S$. These steady states are therefore:

$$q_1 = q_1^*$$
$$V_1 = V_1^*$$
$$r_1 = Cc_1.$$

Since $c_1 \geq 0$, $c_1 \neq 0$ (semi-positive) and $C = (E - A)^{-1} > 0$ (8.8) we have $r_1 > 0$, and from $Cc_1 < \bar{r}$, (36) we have $r_1 < \bar{r}$. Hence (q_1, V_1, r_1) is a point in $\overset{\circ}{\mathcal{X}}$. Conversely, if c_1 violates **36** then r_1 violates $r < \bar{r}$, which means incompatibility.

On the other hand we claim that a point of $\overset{\circ}{\mathcal{X}}{}^S$ is a compatible state if and only if r satisfies

$$(E - A)r \geq 0,$$

i.e. the set of compatible states is:

38 $\quad \mathcal{X}_1^S := \{q_1, V_1, r_1 \mid q_1 > 0, V_1 > {}^{r\ell}v0, 0 < r_1 < \bar{r}, (E-A)r_1 \geq 0\}$

The proof of this statement is straightforward and left to the reader. (Hint: put $c_1 = (E - A)r_1$)

All that remains is to satisfy condition **7.54 e)**, the compatibility of the initial state. We should have $(q_0, V_0, r_0) \in \mathcal{X}_1^S$, i.e.

39 $\quad\quad\quad\quad q_0 > 0, \quad V_0 > {}^{r\ell}v0, \quad 0 < r_0 < \bar{r}$

40 $\quad\quad\quad\quad (E - A)r_0 \geq 0.$

Condition **39** says simply that the initial state must be in the interior of the viability set. **40** stipulates the initial production to yield semi-positive net output. We have hereby proved the following theorem.

41 *Viability theorem for Model S*

Consider the system consisting of the controlled subsystem **8.11-12** and controller **4-5**. (Model S).

Let inputs be admitted such that the output stock norms are

positive, the input stock norms are rℓv-positive, and the end use is semi-positive.

Let state trajectories be viable such that the input stocks, the output stocks and the production are non-negative. Furthermore the production vector does not exceed a given positive capacity constraint $\bar{r}$. (Return transfers are permitted.)

Let the following conditions hold:

a) The system is stable (Theorem **23**).

b) The initial output stocks are positive, the initial input stocks rℓv-positive, and the initial net output is semi-positive (**40**).

The system will then have viable solutions generated by variable end use close enough to the initial net output, initial (input and output) stocks close enough to their norms.

42 Conclusions

From the analysis of Model S we can conclude that it is possible to apply such a controller to the open Leontief-economy (with constant technology) so that the system shows the following characteristics:

a) The producers decide on the production level and on material purchases using information solely on input and output stocks and their rate of change. The decisions are *not coordinated* either between the producers or by any outside control organization, and furthermore they *do not communicate* any information between themselves or to others. (Vegetative non-communicative control.)

b) The decision makers comply with certain simple rules, in which rates of production level and material purchases depend linearly on *deviations of stocks from their given norms* and from the rate of change of these stocks. (PI type control, control by norm principle.)

c) The system is *stable* with appropriately chosen control parameters for any initial state and in this case the trajectories tend towards a *non-Walrasian equilibrium state* for any constant end use.

d) If the system is stable, it is also *viable* in the sense that production and stocks do not turn negative and production never exceeds the given productive capacity. This is valid not only for certain constant end uses within the capacity constraints, but also for variable end uses close enough to such a constant one.

e) The model can be interpreted as a simplified representation of a *buyer's market*.

Chapter 10

THE ORDER SIGNAL
(MODEL B)

1 *Production for stock vs production to order*

When constructing Model S we started with the assumption that the level of output stocks is an important signal on which production decisions are based. This assumption is both logically plausible and empirically supported. However, it is also known that this approach is narrow, order signals play as large a role — if not larger — in production decisions as do output stock signals. (Recall that in 3.9 the observed effect of order signals was more significant than that of stock signals.) There are many reasons why a firm produces for stock or to order or both. Even if we disregard nonstorable commodities (which need not imply production to order, but only that unsold products are destroyed) we must consider at least three aspects.

> (i) There are certain kinds of products which are not usually produced for stock, in order that the special wishes of the buyer may be complied with, or for some other reason. The product is produced according to the delivery time, quantity and quality specified by the buyer (e.g. ship building, clothing made to measure).
>
> (ii) Even for standardized mass-products, it may happen that stocks run short, the buyers are compelled to wait, transfers are decided not by the buyers but by the sellers.
>
> (iii) The product could be produced for stock but it is more convenient both for the seller and the buyer to fix in advance the quantities to be delivered at future dates.

Considering the economy as a whole, production for stock and to order occur in different combinations. Some products are always produced for stock, some only to order and some others to both. For the same product in one period the output stock will pile up, in another period the stock will run out and a backlog of orders arise. In most periods there may be both stocks and backlogs due to market imperfections.

In the first two modeling exercises we confine ourselves to the extreme cases. With Model S we studied the extreme case whereby all the producers always had positive output stocks so that any demand could be satisfied at once. Now we switch to the other extreme; no producer will ever keep output stock, end use will be satisfied from current production, the buyers of input materials will wait until their orders are filled. A backlog of unfilled orders will prevail. The model corresponding to this situation is based on order signal and is called Model B (B for backlog).

2 *Assumption*: The producers do not keep output stocks:

$$q(t) = 0 \qquad \forall t \geq 0.$$

3 *The real processes*

The real processes remain the same as in 8.10 but in view of Assumption 2 equation 8.11 reduces to the algebraic equation

4 $$r = Y\mathbf{e} + c,$$

while the input balance equation 8.12 remains unchanged:

5 $$\dot{V} = Y - \mathbf{A}\langle r \rangle.$$

6 *Notation*

The placing of orders and the bookkeeping of the backlog of unfilled orders, the latter compared to the normal level of backlog orders, will be needed in the build-up of the controller.

$W(t)$: the matrix of order placements ($W^{ij} :=$ the order sent by producer j to the producer of material i).

$K(t) :=$ the matrix of the backlog of orders ($K^{ij} :=$ the accumulated sum of unfilled orders for which producer i is in arrears to producer j.

$K^* :=$ the matrix of the normal stock of backlog orders (K^{*ij} : the normal level of backlog for K^{ij}).

W, K and K* are all n by n matrices.

7 The bookkeeping of orders

The following equation is a bookkeeping identity which by virtue of its economic interpretation belongs to the control sphere of the economy but in the formal analysis will be considered a part of the controlled subsystem. (Cf. 5.7 concerning this discrepancy in the classification.)

8
$$\dot{K} = W - Y$$

i.e. the backlog of orders grows with the amount of incoming new orders and decreases with that of the effected deliveries.

9 The behavioural equations of Model B

In Model B there are two matrices of control variables, one of which is the manipulating matrix Y of transfers, the other the interim control signal W of order placements. We start with the behavioural equation determining W.

CONTROL OF THE ORDER PLACEMENTS

10
$$\dot{W} = -2\beta\gamma\dot{V} - \gamma^2(V - V^*)$$

This equation is similar to equation 9.5, that which controlled the transfer in Model S. Here the placement of orders is controlled by the buyer in the same way (depending on input stock variables and norms) as transfers were controlled by him there.

The transfers will not now be decided by the buyers but by the suppliers, taking into account the corresponding backlog of orders.

CONTROL OF TRANSFERS

11
$$\dot{Y} = 2\beta\gamma\dot{K} + \gamma^2(K - K^*).$$

The structure of this equation is similar to that of the former behavioural equations; notice, however, the change of sign on the right hand side. This indicates a positive feedback instead of the negative ones applied so far. Nonetheless, it is, of course, reasonable to assume that both order backlogs above the normal level and the growth of the backlogs induce suppliers to deliver more.

12 *The end use*

It may seem unreasonable, however, to deal with end users differently from productive users. Producers have to place orders and wait for their fulfillment, whereas a buyer for end use is immediately satisfied from current production *as if* the latter enjoyed a priority. In fact this priority assumption is not needed here. Rather we can observe that it is immaterial to the operation of the system whether the transfer for the exogenous end use is given directly, or whether the placement of orders is given in this way and the transfer is a function of the orders. In fact we could add two equations to the system in analogy to 8 and 11 referring to the bookkeeping of backlog orders from, and transfers to, the end user. This pair of equations could be solved independently of the rest of the system and would add nothing to the analysis. This is why they have been omitted.

In other words, the end use c represents at the same time the order sent to the supplier by the end user and the transfer of commodity from the supplier to the consumer, since end use is satisfied immediately.

Alternatively, we could assume that the initial ($t = 0$) backlog of orders from the end user equals its norm for each commodity and the initial placement of orders by him equals the initial transfer. This amounts to saying that the end user starts in equilibrium state and remains there permanently.

Whichever way we go round the problem the priority assumption need not be made in this model.

13 *Interpretation of the behavioural equations, structural analysis and market relations*

Let us start with the placement of orders. This is decided by the buyer according to equation **10** in the knowledge of his own input stocks. He can observe them in an *introspective* way and make an *uncoordinated* decision. But the placement of orders is a *transactionally communicative* process since the result of this decision is not a manipulating variable but a signal; an intermediate control variable to be transmitted to the supplier.

The supplier i receives the orders and faces the task of deciding on the transfers Y^{ij}. From equations **8** and **11**

$$\dot{K}^{ij} = W^{ij} - Y^{ij}$$

$$\dot{Y}^{ij} = 2\beta\gamma\dot{K}^{ij} + \gamma^2(K^{ij} - K^{*ij}).$$

Let us not concern ourselves with the method by which the supplier solves this differential equation system in Y^{ij} and K^{ij} (e.g. by trial and error, or approximating the differential equations by difference equations.) From the point of view of structural analysis the relevant point is that the supplier has all the information (W^{ij} and K^{*ij}) necessary to decide on Y^{ij}, therefore this decision needs *no coordination*. The decision then enters his order book (*non-communicative* signal transmission) and is effected by the delivery made by himself (introactive manipulation).

Finally the decision on production is made simply by summing up the transfers thus decided and by adding to this the demand c of the end user which has also been communicated to the supplier.

Hence Model B represents a *vegetative* control structure with *transactional communication* (Second stage, 4.14). Figure 6 depicts the block diagram of Model B. It shows clearly the double-loop structure: the small positive feedback loop within the large negative one.

The *market* represented by Model B is very different from that in Model S. Both represent orderly markets where the short side rule applies. But here it is the suppliers who are always on the short side and the buyers are — so to speak — "rationed" by the suppliers' propensity to produce and transfer (except the buyers for end use who are not rationed). This is therefore essentially a seller's market.

Figure 6

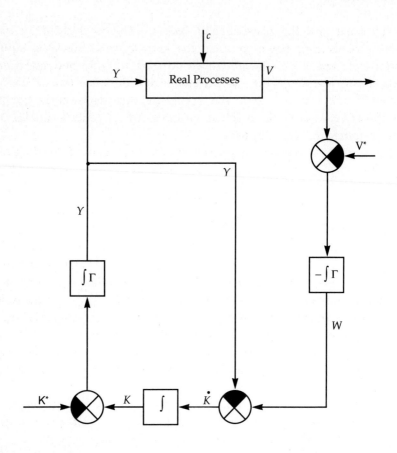

Block diagram of Model B

14 Stability analysis: transformation to the reduced form

We will start again by transforming the system to the *reduced form* 8.24-25 by aggregation.

Let us post-multiply equations 5, 8, 10 and 11 by e, substitute $r = Ye + c$ from 4 to the aggregated 5, and multiply the aggregated 8 by (−1). In this way we obtain the following system

15 $$\dot{V}e = (E - A)Ye - Ac$$

16 $$-\dot{K}e = -We + Ye$$

17 $$\dot{W}e = -2\beta\gamma(\dot{V}e) - \gamma^2(V - V^*)e,$$

18 $$\dot{Y}e = -2\beta\gamma(-\dot{K}e) - \gamma^2(-K + K^*)e,$$

where **15-16** is the controlled subsystem and **17-18** the controller. With the notation

19 $$y = \begin{bmatrix} Ve \\ -Ke \end{bmatrix}, \quad v = \begin{bmatrix} We \\ Ye \end{bmatrix}, \quad z = \begin{bmatrix} -Ac \\ 0 \end{bmatrix}, \quad y^* = \begin{bmatrix} V^*e \\ -K^*e \end{bmatrix}$$

20 $$Q^B = \begin{bmatrix} 0 & E - A \\ -E & E \end{bmatrix}$$

we obtain **15-18** in the reduced form.

21 Stability of the reduced form

In order to establish stability of the reduced form we will give the conditions in terms of the spectrum $\mathcal{F}$ of matrix **A** for Q^B being simple and having all eigenvalues with positive real part. These conditions are given in the form of a lemma. Let us denote by $\mathcal{Q}^B$ the spectrum of Q^B.

22 Lemma

Under Assumptions 8.7

a) Q^B is simple if $0.75 \notin \mathcal{F}$.

b) Re $\sigma > 0$ for all $\sigma \in \mathcal{Q}^B$ if and only if

23
$$\text{Re } \varphi + (\text{Im } \varphi)^2 < 1 \qquad \forall \varphi \in \mathcal{F}.$$

c) **23** holds for some $\varphi \in \mathcal{F}$ whenever Im $\varphi = 0$, or Re $\varphi \leq 0$, or

24
$$|\varphi| < \frac{\sqrt{3}}{2} \approx 0.866.$$

25 Proof of Lemma 22

Applying the same trick as in 9.**20** we obtain the characteristic equation for Q^B in the form:

26
$$\sigma^2 - \sigma + 1 = \varphi \qquad \forall \varphi \in \mathcal{F}.$$

Solving equation **26** for σ we obtain

$$\sigma = \tfrac{1}{2}(1 \pm \sqrt{4\varphi - 3}).$$

a) Equation **26** has two different roots unless $\varphi = 0.75$, which we have excluded. We cannot have equal σ's for different φ's, and A had n different eigenvalues. Thus Q^B has 2n different eigenvalues and is hence simple.

b) We want to have

$$\text{Re } 2\sigma = 1 \pm \text{Re } \sqrt{4\varphi - 3} > 0,$$

i.e.

27
$$\pm \text{Re } \sqrt{4\varphi - 3} < 1.$$

By formula 8.38

$$\text{Re } \sqrt{4\varphi - 3} = \sqrt{\tfrac{1}{2}(|4\varphi - 3| + 4\text{Re}\varphi - 3)}$$

and squaring inequality **27** we obtain

$$|4\varphi - 3| < 5 - 4\text{Re}\varphi.$$

The right hand side is positive, thus we square again:

$$(3 - 4\text{Re}\varphi)^2 + (4\text{Im}\varphi)^2 < (5 - 4\text{Re}\varphi)^2,$$

which after rearrangement gives **23**.

c) If $\text{Im}\varphi = 0$ or $\text{Re}\varphi \leq 0$ then **23** holds since $|\varphi| < 1$. Furthermore, if $|\varphi| < \sqrt{3}/2$, then

$$\text{Re}\varphi + (\text{Im}\varphi)^2 < \text{Re}\varphi + \tfrac{3}{4} - (\text{Re}\varphi)^2 = 1 - (\tfrac{1}{2} - \text{Re}\varphi)^2 \leq 1.$$

28 Discussion of condition 23

Condition **23** is not empty. For instance, $\varphi = 0.5 + 0.72\iota$ can belong to $\mathcal{F}$ since $|\varphi| = 0.8766 < 1$, but $\text{Re}\varphi + (\text{Im}\varphi)^2 = 1.0184 > 1$ which violates **23**.

The implications of inequality **23** can best be shown by a geometrical representation on the complex number plane. In Figure 7 the radius of the heavy circle is $\rho < 1$, so that the full spectrum of **A** is in this circle, the area to the left of the heavy line parabola is the locus of the eigenvalues satisfying **23**. The simply shaded region is thus the locus of the eigenvalues satisfying both $|\varphi| \leq \rho$ and **23**, thus the eigenvalues which are admissible, and the tiny doubly shaded region between the circle and the parabola (which is empty if $\rho < \sqrt{3}/2$) is the locus of the destabilizing eigenvalues of **A**.

Thus we are faced with the following strange situation. The technology may be characterized by such a coefficient matrix that the system cannot be stabilized by any choice of the control parameters if the behavioural rules are those of Model B. In other words: *stabilizability depends on technology*. This is a phenomenon I have not encountered in any other model I have been dealing with, and it is the more embarrassing since I have been unable to read any economic meaning into condition **23**.

122   CHAPTER 10

Figure 7

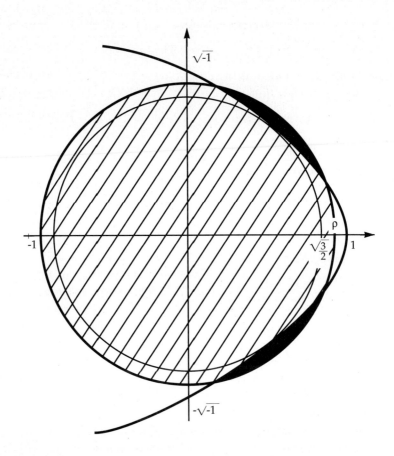

Locus of the destabilizing eigenvalues of **A**
in Model B

29 Empirical study of condition 23

The question arises, however, that even if eigenvalues violating 23 cannot be *theoretically* excluded, what is the matter with empiria? Do actual input coefficient matrices really have eigenvalues violating 23? I found only two input coefficient tables with the whole spectrum made available. One is an 18 sector table of the USA for 1939 (MORGENSTERN, 1954, p. 413), the other is a 12 sector table of Japan (TSUKUI, 1968). Table 3 summarizes the data relevant to our inquiry.

Table 3

	USA	Japan
Spectral radius (ρ)	$\varphi_1 = 0.5415$	$\varphi_1 = 0.919$
The largest (in modulus) pair of complex eigenvalues with negative real part	$\varphi_{2,3} = -0.1968 \pm 0.0871\iota$ $\|\varphi_{2,3}\| = 0.2153$	— —
The largest (in modulus) pair of complex eigenvalues with positive real part	$\varphi_{16,17} = 0.0155 \pm 0.0069\iota$ $\|\varphi_{16,17}\| = 0.0170$	$\varphi_{7,8} = 0.158 \pm 0.053\iota$ $\|\varphi_{7,8}\| = 0.167$
The value of the l.h.s. of **23** for these latter eigenvalues [Re φ + (Im $\varphi)^2$]	0.0155	0.161

The "dangerous" eigenvalues are in both cases very far from violating **23**, what is more their modulus is far less than the critical value $\sqrt{3}/2$. If I were not to be called names for drawing conjecture from two cases, I would dare to suggest that complex eigenvalues of input coefficient matrices drawn from real data are small relative to their spectral radius. If this conjecture turned out to be valid in many cases, it would be worth seeking the structural characteristics of these tables which exclude complex eigenvalues close to the boundary of the spectral circle. A reasoning inspired by the results of OSTROWSKI (1974) leads me to presume that this phenomenon may be related to the fact that the *entries* of such matrices are small relative to ρ. The formulae of Ostrowski yield estimations (not for the complex but for the non-dominant eigenvalues, to wit) which were not powerful enough to settle my question.

30 Stability analysis of the original system

By Lemma **22** we have specified sufficient conditions on **A** under which the reduced form of **15-18** can be stabilized by choosing β according to 8.31 b). If the reduced form is stable in v it is also stable in $Ye = [0\ E]v$ and hence in $r = Ye + c$. But we still have to check the stability for the individual components V^{ij}, Y^{ij}, W^{ij}, K^{ij}. For this let us select the i,j components from equations **5, 8, 10** and **11** to obtain an equation system in the four variables just mentioned:

$$\dot{V}^{ij} = Y^{ij} - \mathbf{A}^{ij}r^j$$

$$-K^{ij} = -W^{ij} + Y^{ij}$$

$$\dot{W}^{ij} = -2\beta\gamma\dot{V}^{ij} - \gamma^2(V^{ij} - V^{*ij})$$

$$\dot{Y}^{ij} = -2\beta\gamma(-\dot{K}^{ij}) - \gamma^2(-K^{ij} + K^{*ij}).$$

Applying the notation

$$y = \begin{bmatrix} V^{ij} \\ -K^{ij} \end{bmatrix}, \quad v = \begin{bmatrix} W^{ij} \\ Y^{ij} \end{bmatrix}, \quad z = \begin{bmatrix} -\mathbf{A}^{ij}r^j \\ 0 \end{bmatrix}, \quad y^* = \begin{bmatrix} V^{*ij} \\ -K^{*ij} \end{bmatrix}$$

$$\bar{Q}^B = \begin{bmatrix} 0 & 1 \\ -1 & 1 \end{bmatrix},$$

we have again a reduced form at hand. The eigenvalues of $\bar{Q}^B$ are

$$\bar{\sigma}_{1,2} = \tfrac{1}{2}(1 \pm \iota\sqrt{3}\,)$$

which have positive real part and

$$\vartheta(\bar{\sigma}_{1,2}) = \frac{\sqrt{6}}{4} \sim 0.6124 \ .$$

Thus $\beta > 0.6124$ has to be added to the stability conditions. Hereby we have proved:

31 *Stability theorem for Model B*

If the input coefficient matrix **A** (with spectrum $\mathcal{F}$) satisfies Assumptions **8.7** and additionally

(i) $0.75 \notin \mathcal{F}$

(ii) $\mathrm{Re} \ \varphi + (\mathrm{Im} \ \varphi)^2 < 1$ $\quad \forall \varphi \in \mathcal{F}$,

then the Model S consisting of the controlled subsystem **4, 5, 8** and the controller **10-11** is stable if and only if

$$\beta > \max \{ \frac{\sqrt{6}}{4}, \max \{\vartheta(\sigma) | \sigma \in \mathcal{Q}^B\}\}$$

with $\mathcal{Q}^B$ defined as the spectrum of $\mathbf{Q}^B$ from **20**.

32 *The constant-input steady state*

Following the procedure we applied to Model S in 9.**25**, for the constant end use $c(t) = c$ we can calculate the following steady (= equilibrium) state for Model B if it is stable:

33 $\quad\quad\quad\quad V_e = V^*, \quad\quad K_e = K^*$

34 $\quad\quad\quad\quad r_e = Cc, \quad\quad Y_e = W_e = A<Cc>,$

and the interpretation of these components of the equilibrium point corresponds to what was ascertained under 9.**25**.

The *non-Walrasian* character of this equilibrium state is inferred from the perpetuation of positive input stocks and backlog of orders, which are not needed for the realization of equilibrium production and transfer.

35 *Viability analysis*

In the viability analysis of Model B we follow the same line of reasoning outlined in 9.**30** for Model S. In order to ease the notational burden I shall not point out that some positivity requirements refer to the *relevant* entries of Y, V and K only. I hope that the reader is already sufficiently familiar with this distinction and can fill in the incomplete notation without notes.

36 *Definition of the viability set and the admissible input set*

37 $$\mathcal{X}^B := \{V,\ K,\ Y,\ r \mid V{\geq}0,\ K{\geq}0,\ Y{\geq}0,\ r{\leq}\bar{r}\},$$

where $\bar{r}$, as earlier, is the constant positive capacity vector.

38 $$\mathcal{U}^B := \{V^*,\ K^*,\ c \mid V^*{>}0,\ K^*{>}0,\ c{\geq}0\}.$$

(The condition $c \neq 0$ is omitted here and will be omitted in the sequel for sake of brevity.)

39 *Interpretation of the viability conditions*

The input stock V and the backlog of orders K must be non-negative by their economic interpretation.

By $Y \geq 0$ we excluded return transfer of commodities, which was admitted in Model S. But here, where transfers are decided by the supplier, the return transfer would mean that the supplier demanded return of his product and the user complied with this demand in spite of his outstanding orders for the product in question. Thus it would be meaningless not to require non-negativity of the transfers. The placement of orders W, however, is not restricted to be non-negative: outstanding orders can be withdrawn by the user.

$r \geq 0$ need not be stipulated, since from **4** and from $Y \geq 0$ it follows that:

40 $$r = Y\mathbf{e} + c \geq c \geq 0,$$

which guarantees not only that r is non-negative but also that $r \geq c$, i.e. that the demand of the end user really can be satisfied from the current production, as was assumed (but not hitherto formalized).

41 *Compatibility*

Following the line of reasoning applied in 9.35 we can simply write down the sets of compatible inputs and states:

42 $$\mathcal{U}_1^B := \{V_1^*, K_1^*, c_1 | V_1^*>0, K_1^*>0, c_1 \geq 0, Cc_1 < \bar{r}\}$$

43 $$\mathcal{X}_1^B := \{V_1, K_1, Y_1, r_1 | V_1>0, K_1>0, Y_1>0, r_1<\bar{r}, (E-A)r_1 \geq 0\}.$$

In this way the initial state (V_0, K_0, Y_0, r_0) is compatible if and only if

44 $$V_0 > 0, \quad K_0 > 0, \quad Y_0 > 0, \quad r_0 < \bar{r}$$

45 $$(E - A)r_0 \geq 0.$$

46 *Viability Theorem for Model B*

Consider the system consisting of the controlled subsystem **4, 5, 8** and controller **10-11** (Model B). Let inputs be admitted such that the input stock, norms and the norms of order backlog are $r\ell v$-positive, the end use is semi-positive.

Let trajectories be viable such that the input stocks, the stocks of backlog order and the transfers are non-negative, the production vector does not exceed a given capacity constraint $\bar{r}$. (Withdrawal of orders is permitted.)

Let the following conditions hold:

a) The system is stable (Theorem **31**)

b) The initial input stocks, the initial backlog of orders and the initial transfers are $r\ell v$-positive, the initial net output is semi-positive (**44, 45**).

Then the system has viable solutions generated by variable end use.

47 *Conclusions*

From the analysis of Model B we draw the conclusion that to the same real sphere as that of Model S we can fit another controller, such that the system shows the following characteristics:

a) The producers decide on the production level, furthermore, as buyers, they decide on the placement of orders and, as suppliers, on the transfers. All their decisions are *uncoordinated*. There is *transactional communication* between pairs of buyers and suppliers in the form of orders flowing from the buyer to the supplier. It is important to notice that the transmitted signal is of a quantity (non-price) character. (Vegetative control with transactional quantity communication.)

b) The decision makers comply with some simple behavioural rules: they do not keep output stocks, change the placement of orders and the transfers taken as functions of the deviation of input stocks and backlog of orders, respectively, from their norms and of the derivatives of these arguments. (PI type control by norm.)

c) It may happen that this control makes the system unstable for any choice of the control parameters depending on the numerical structure of the technology matrix **A**. It can nevertheless be expected that such destabilizing sets of technologies will hardly occur in practice. If the input coefficient matrix is not so awkward, the system *can be stabilized* by appropriately chosen control parameters and the trajectories then tend to a *non-Walrasian equilibrium* state for any constant end use.

d) If the system is stable, then it is also *viable*, i.e. the trajectories satisfy the reasonable non-negativity requirements, and the production can be kept within the given capacity constraints (under restrictions imposed upon the variable end use).

e) The model can be interpreted as a simplified representation of a *seller's market*.

Chapter 11

THE COMBINATION OF STOCK AND ORDER SIGNALS
(MODEL SB)

1 *Motivation*

In the preceding two chapters I dealt with two extreme cases; in Model S there was no backlog of orders and the suppliers kept output stocks all the time; in Model B they kept no output stock but a backlog of orders prevailed. Both represented orderly markets. To extend the study to orderless markets (Cf. 2.15), *where there are both output stocks and backlog of orders*, is not only the logical next step to be taken, but is also supported by the empirical evidence discussed in Chapter 3, which showed that the firms reported this occurrence in 83 percent of the observed cases.

2 *The real processes*

The real processes are invariably represented by the well known equations:

3 $$\dot{q} = r - Ye - c$$
4 $$\dot{V} = Y - A<r>.$$

5 *The control processes*

THE BOOKKEEPING OF BACKLOG ORDERS

6 $$\dot{K} = W - Y$$

THE CONTROL OF PRODUCTION

7
$$\dot{r} = -2\beta\gamma\dot{q} - \gamma^2(q - q^*)$$

THE CONTROL OF THE PLACEMENT OF ORDERS

8
$$\dot{W} = -2\beta\gamma\dot{V} - \gamma^2(V - V^*)$$

THE CONTROL OF TRANSFERS

9
$$\dot{Y} = 2\beta\gamma\dot{K} + \gamma^2(K - K^*)$$

These four equations concur with equations 10.8, 9.4, 10.10 and 10.11 in turn. Their interpretation and structural analysis is also the same and will be briefly abstracted here.

10 *Interpretation of the controller*

The producers decide on production levels in view of their output stocks (7) and as buyers decide on the placement of orders considering their input stocks (8). The suppliers decide on the transfers on the basis of the backlog of orders (9) which is recorded by them (6). 9 and 6 are solved as a simultaneous system alike for Model B. The demand for end use is immediately satisfied *ex store*. The alternative explanations offered in 10.12 for the disparity between the handling of demand for end and for productive use is valid here also.

11 *Structural analysis. The orderless market*

Since Model SB combines the decision and information processes of Model B and Model S, its structural analysis would consist of repeating what was said there. I will therefore content myself with the conclusion: the controller is *vegetative* with *transactional* (quantity) *communication* among the producers.

The control structure is shown also in Figure 8, consisting of a single and a double loop working in parallel.

It is the *market structure* represented by Model SB which deserves special attention, being — as I mentioned — *orderless*. Let us recall that in an orderly market either all the buyers can realize their intention to buy or all the sellers their intention to sell in the market for a

Figure 8

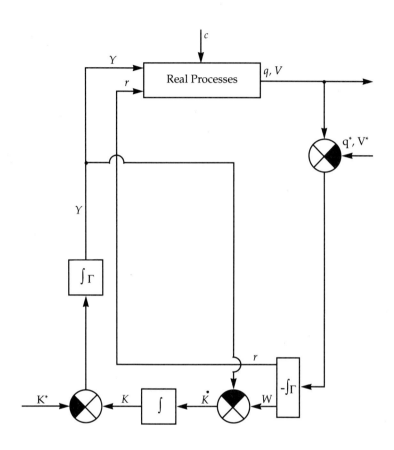

Block diagram of Model SB

given commodity. It can never occur that neither of them realizes their intentions.

Among the variables of Model SB there is none which could be called "intention to sell" or "intention to buy". However, with certain reservations, the output stocks can be considered as representing the intention to sell, since the commodities were produced for sale, and the backlog of orders as representing intention to buy, since the orders were transmitted just to communicate this intention to the supplier.

We see that in Model SB the producers as sellers cannot realize their intention as long as output stocks are positive, neither are they satisfied as buyers since they have to wait for the fulfillment of the orders. Hence the *short-side rule fails to apply*, no short side or long side develops on this market. If you wish, everybody feels as if they are on the long side — except for the buyers for end use who are always satisfied from the output stock immediately.

12 *Stability analysis: transformation to the reduced form*

Let us post-multiply equations **4, 6, 8** and **9** by **e** to get the reduced form equations 8.**24-25** with the following notation:

13
$$y = \begin{bmatrix} q \\ Ve \\ -Ke \end{bmatrix}, \quad v = \begin{bmatrix} r \\ We \\ Ye \end{bmatrix}, \quad z = \begin{bmatrix} -c \\ 0 \\ 0 \end{bmatrix}, \quad y^* = \begin{bmatrix} q^* \\ V^*e \\ -K^*e \end{bmatrix}$$

14
$$Q^{SB} := \begin{bmatrix} E & 0 & -E \\ -A & 0 & E \\ 0 & -E & E \end{bmatrix}$$

As earlier we introduce the stability conditions in the form of a lemma.

15 *Lemma*

Under Assumptions 8.7

a) Q^{SB} is simple if $\frac{1}{27}(7 \pm 14\sqrt{2}) \notin \mathcal{F}$.

b) Re $\sigma > 0$ for all $\sigma \in \mathcal{Q}$.

16 Proof of lemma 15

As before (9.20) we obtain the characteristic equation of Q^{SB}:

17
$$-\sigma^3 + 2\sigma^2 - 2\sigma + 1 = \varphi \quad \forall \varphi \in \mathcal{F}$$

a) Equation **17** has three different roots unless the derivative of the left hand side of **17** disappears:

$$-3\sigma^2 + 4\sigma - 2 = 0,$$

which gives $\sigma = \frac{1}{3}(2\pm\imath\sqrt{2})$, $\varphi = \frac{1}{27}(7\pm\imath 4\sqrt{2})$, the excluded eigenvalues of A. Equal σ's cannot give unequal φ's, hence Q^{SB} has 3n different eigenvalues and is simple.

b) By the use of the notation (8.35)

$$\sigma = \mu + \imath\upsilon$$

(μ and υ are real) we separate the real and the imaginary part of equation **17**:

$$\text{Re } \varphi = (3\mu - 2)\upsilon^2 + (-\mu^3 + 2\mu^2 - 2\mu + 1)$$

$$\text{Im } \varphi = \upsilon^3 - \upsilon(3\mu^2 - 4\mu + 2).$$

From these we obtain:

18
$$|\varphi|^2 = \upsilon^6 + (3\mu^2 - 4\mu)\upsilon^4 + (3\mu^4 - 8\mu^3 + 8\mu^2 - 2\mu)\upsilon^2 + (-\mu^3 + 2\mu^2 - 2\mu + 1)^2$$

and we have to satisfy the inequality $|\varphi|^2 \leq \rho^2 < 1$. Observe that on the right hand side of **18** the odd powers of μ have all negative signs. For $\mu \leq 0$ we thus would have four non-negative terms of which the last one is not less than one. Thus we could not have $|\varphi| < 1$ for $\mu \leq 0$, by contradiction we conclude $\mu = \text{Re } \sigma > 0$.

19 Stability analysis of the original system

From lemma **15** and from theorem 8.31 it follows that the reduced form is stable (under condition **15** a)) for β satisfying 8.31 a). If the reduced form is stable in y and v then the original system is stable in

$$r = [E, 0, 0]v$$

and

$$q = [E, 0, 0]y.$$

For the rest of the variables we obtain from **4, 6, 8, 9** the same system of four scalar equations which we analysed in **10.30**, and which proved to be stable if $\beta > \sqrt{6}/4$. Hereby we have proved:

20 *Stability theorem for Model SB*

If the input coefficient matrix **A** (with spectrum $\mathcal{F}$) satisfies Assumptions **8.7** and additionally

$$\tfrac{1}{27}(7 \pm 14\sqrt{2}) \notin \mathcal{F}$$

then the Model SB consisting of the controlled subsystem **3, 4, 6** and controller **7-9** is stable if and only if

$$\beta > \max \left\{ \frac{\sqrt{6}}{4}, \max \{\vartheta(\sigma) \mid \sigma \in \mathcal{Q}^{SB}\} \right\}.$$

where $\mathcal{Q}^{SB}$ is the spectrum of matrix Q^{SB} in **14**.

21 *The constant-input steady state*

For the constant end use **c** we calculate the constant input steady state for a stable Model SB in the familiar way (cf. **10.32**):

22 $$q_e = q^*, \quad V_e = V^*, \quad K_e = K^*$$

23 $$r_e = Cc,$$

24 $$W_e = Y_e = A<Cc>.$$

This equilibrium state is *non-Walrasian*, in fact less Walrasian than the previous two. (The Walrasian equilibrium would be characterized by $q_e = 0$ *and* $K_e = 0$. In Model S and Model B we had $q_e > 0$ *or* $K_e > 0$ respectively, here we will have both positive.)

25 Viability analysis

The viability analysis follows the familiar route of defining the viability set $\mathcal{X}^{SB}$, the admissible input set $\mathcal{U}^{SB}$, the compatible input set $\mathcal{U}^{SB}_1$ and the set of compatible states $\mathcal{X}^{SB}_1$. The initial state is then required to be compatible.

26 $$\mathcal{X}^{SB} := \{q, V, K, Y, r \mid q \geq 0, V \geq 0, K \geq 0, Y \geq 0, 0 \leq r \leq \bar{r}\}$$

27 $$\mathcal{U}^{SB} := \{q^*, V^*, K^*, c \mid q^* > 0, V^* > 0, K^* > 0, c \geq 0\}$$

28 $$\mathcal{U}^{SB}_1 := \{(q^*_1, V^*_1, K^*_1, c_1) \in \mathcal{U}^{SB} \mid Cc < \bar{r}\}$$

29 $$\mathcal{X}^{SB}_1 := \{(q_1, V_1, K_1, Y_1, r_1) \in \overset{\circ}{\mathcal{X}}{}^{SB} \mid (E-A)r_1 \geq 0\}.$$

We can see once more that the interior of $\mathcal{X}^{SB}$ is non-empty as is $\mathcal{U}^{SB}_1$. The interpretation of the viability criteria is the same as before, return transfer is not permitted, while withdrawal of orders is.

Viable solutions exist if the initial state belongs to $\mathcal{X}^{SB}_1$.

30 Viability theorem for Model SB

Model SB (**3, 4, 6-9**) has viable solutions in terms of the viability set $\mathcal{X}^{SB}$ and admissible input set $\mathcal{U}^{SB}$ (**26, 27**) assuming stability (Theorem 20) and compatible initial state that belongs to $\mathcal{X}^{SB}_1$ (**29**). Viable solutions are generated also by variable end use. (A more explicit viability theorem can be formulated by analogy to Theorems 9.**41** and 10.**46**.)

31 Conclusions

We have fitted to the open Leontief economy another controller which was a combination of the controllers of the stock signal (S) and the order signal (B) models. All its characteristics (vegetative control with transactional (quantity) communication, PI type controller, stability, non-Walrasian equilibrium, viability) coincided with those of Model B with the exception of two particular properties:

a) As in Model S but unlike Model B the system was stabilizable for any input coefficient matrix **A** (excluding some specific values from the spectrum of **A**).

b) The system could be interpreted as the representation of an orderless market, where the "short side rule" did not apply.

Chapter 12

THE COMMERCIAL STOCK SIGNAL (MODEL C)

1 *Introduction of the commercial sector*

The first three models were constructed on the tacit assumption that transfer of a given commodity is made only from the producer of this commodity to the user for productive or end use. In real life there are two more kinds of transfer which have been assumed away hitherto:

(i) There may be a transfer of a commodity between producers (as well as end users) such that neither of the parties is a producer of that commodity. For example, a producer has an excess input stock of some material which he sells to another user. Such processes will not be modelled in the future either.

(ii) There are organizations, called *traders*, who deal exclusively with commodity transfers. The set of all traders forms the *commercial sector*. The distinctive role of the commercial sector will be studied in Model C (C for commercial) in a simplified setup.

The construction of such a model was also motivated by my uneasiness that in most theoretical models aiming at a description of the functioning of the market, the trader, the leading character in all developed markets, does not appear on the scene as such, or plays only second fiddle and is dispensable. I am not claiming that this theoretical gap is filled in by the following model or that the indispensability of commerce is demonstrated by it, but at least it shows how naturally the commercial sector fits into the control of the turnover of goods.

2 Assumptions on the commercial sector

In the modelling of the economy with a separate commercial sector the following simplifying assumptions will be made:

a) There is only one commercial firm: the *trader*.

b) All the commodity transfers go through the trader; in each case he is either the buyer or the seller.

Assumption a) could be easily substituted by the assumption that there is a separate trader for each commodity (e.g. the sales department of the producing firm). This assumption would lead to the same model with slightly different interpretations of the control structure. The more realistic assumption of having more traders for each commodity could not be reconciled with the starting "one good - one producer" assumption.

In real life one part of the transfer goes directly from the producer to the user and another part involves the trader as middleman. Since in the first three models I studied one extreme case of no indirect trade, here we turn to the other extreme of no direct trade by assumption b).

3 Informal interpretation of the real processes

Although the equations describing the real processes will remain formally unchanged, their interpretation and even the definition of some variables will be adjusted to the new situation.

The producers do not keep output stocks but transfer all that is produced to the trader. They thus produce to an order placed by the trader. Hence q, hitherto the producer's output stock, will be reinterpreted as *commercial stock* kept by the trader. At the same time r, the level of production, also represents the transfer from the producers to the trader. The transfer matrix Y and end use c are reinterpreted as transfers from the trader to the producer as buyer of input commodities and to the end user, respectively. Both are served from the commercial stock.

4 The equations of the real processes

THE BALANCE OF COMMERCIAL STOCKS

5
$$\dot{q} = r - Ye - c$$

i.e. the rate of the change of the commercial stock equals the transfer (= production) from the producer to the trader minus the transfer from the trader to the users.

THE BALANCE OF INPUT STOCKS

6
$$\dot{V} = Y - A<r> .$$

Here everything continues along the now familiar route except that the material transfer Y now comes from the trader instead of the producers.

7 Notation

$w(t) :=$ the n-vector of orders placed by the trader with the producers ($w^i :=$ the order for product i).

$k(t) :=$ the n-vector of the backlog of order at the producers ($k^i :=$ the backlog at producer i being owed to the trader)

$k^* :=$ the norm-vector of the backlog.

8 The bookkeeping of the backlog of orders

9
$$\dot{k} = w - r$$

i.e. the rate of change of the backlog is the difference between the order incoming from the trader and the level of production which equals the transfer to the trader. The book of orders is kept with the producer. (Cf. 10.8)

10 *The behavioural equations of Model C*

CONTROL OF ORDER PLACEMENTS

11
$$\dot{w} = -2\beta\gamma\dot{q} - \gamma^2(q - q^*)$$

The placement of orders is decided by the trader as buyer depending on his stocks (and norms) in the usual way.

CONTROL OF TRANSFERS (from the trader to the producers)

12
$$\dot{Y} = -\beta\gamma\dot{V} - \gamma^2(V - V^*).$$

The rule and its interpretation are the same as in Model S (9.5). The transfers are decided by the buyers, but the supplier is the trader, not the producer. The same applies to the transfers for end use; in this model there is no distinction between buyers for productive and for end use.

CONTROL OF PRODUCTION (= transfer to the trader)

13
$$\dot{r} = 2\beta\gamma\dot{k} - \gamma^2(k - k^*).$$

The level of production and thereby the transfer to the trader is decided by the producer on the basis of the backlog of orders, much as in Model B. (Cf. 10.11 and 10.4)

The situation whereby the producer has to solve an equation system consisting of a scalar component of **9** and **13** is similar to Model B.

14 *Structural analysis. The market structure*

Equations **9**, **12** and **13** represent vegetative, non-communicative elements, while **11** is vegetative and transactionally communicative: the trader sends a (quantity) signal to the producer. Hence the controller as a whole is again *vegetative* with transactional (quantity) communication.

In the block diagram of Model C (Figure 9) we see a simple and a double control loop, interconnected with each other.

The *market structure* represented by this model is a peculiar one. On the one hand there is the market where the producers are the sellers and the trader is the buyer. Let us call this the *wholesale*

Figure 9

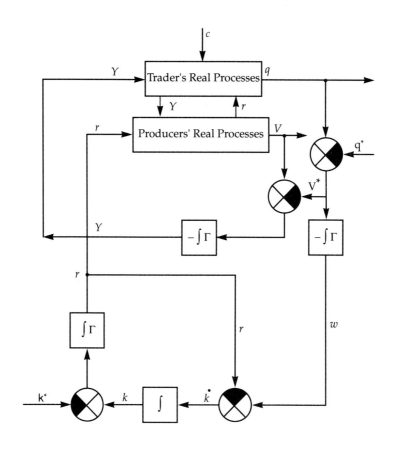

Block diagram of Model C

market. This is an orderly seller's market, with the producers on the short side and the trader on the long side. But there is also another market, the *retail* market, where the trader is the seller and the producers and end users are the buyers. This is an orderly buyer's market, where the trader is on the long side and the buyers on the short side. If, however, we put the two markets together we find both backlog of orders with the producer and unsold stocks with the trader. Hence the united market is orderless. Even so, the effect of the trader is that both the producers as sellers and all the buyers are satisfied, and on both markets the trader is on the long side. Here the *role* of the trader is to absorb the market imperfections, and to make both the wholesale and the retail market orderly. The short side rule, which did not apply to the united market of Model SB, applies now to both separate markets of Model C. This is why I have said that commerce fits naturally in the control of the turnover of goods.

Another effect of the entry of commerce is an economizing in information. In Model SB we had $n + 2n^2 - 2n_0$ norms (q^*, V^*, K^*); in Model C we only have $2n + n^2 - n_0$ of them. The number of signals communicated also falls from n^2 (W) to n (w). Thus the commercial sector — without turning into a control organization separated from real processes — also plays the role of an information gathering and condensing agency.

15 *Stability analysis*

The transformation of equations 5, 6, 9, 11-13 to the reduced form requires now the post-multiplication of equations 6 and 12 by e and the following notation:

16
$$y = \begin{bmatrix} q \\ Ve \\ -k \end{bmatrix}, \quad v = \begin{bmatrix} w \\ Ye \\ r \end{bmatrix}, \quad z = \begin{bmatrix} -c \\ 0 \\ 0 \end{bmatrix}, \quad y^* = \begin{bmatrix} q^* \\ V^*e \\ -k^* \end{bmatrix}$$

17
$$Q^C := \begin{bmatrix} 0 & E & E \\ 0 & E & -A \\ -E & 0 & E \end{bmatrix}.$$

The characteristic equation and thus the spectrum $\mathscr{Q}^C$ of Q^C coincides with $\mathcal{Q}^{SB}$ (see **11.17**); Q^C and Q^{SB} are similar matrices.

Thus Lemma **11.15** is valid if Q^C replaces Q^{SB}, and under its condition the reduced form is stable. From this the stability of the original system follows in the variables q, k, w and r. With respect to the variables V^{ij}, Y^{ij} we have the two scalar equations:

$$\dot{V}^{ij} = Y^{ij} - A^{ij}r^j$$

$$\dot{Y}^{ij} = -2\beta\gamma\dot{V}^{ij} - \gamma^2(V^{ij} - V^{*ij})$$

This system has already been analysed in **9.22**, where it turned out not to require any new condition for stability. Hereby we have proved the following theorem.

18 Stability Theorem for Model C

If the input coefficient matrix A (with spectrum $\mathcal{F}$) satisfies Assumptions **8.7** and additionally

$$\tfrac{1}{27}(7 \pm 14\sqrt{2}\,) \notin \mathcal{F},$$

then the Model C consisting of the controlled subsystem **5, 6, 9** and the controller **11-13** is stable if and only if

$$\beta > \max\, \{\vartheta(\sigma)\, | \,\sigma \in \mathcal{Q}^C\},$$

where $\mathcal{Q}^C$ is the spectrum of matrix Q^C in **17**.

19 The constant-input steady state

If Model C is stable we can calculate the steady (= equilibrium) state for constant end use $c(t) = c$ in the usual way (Cf. **9.25**).

20 $\qquad\qquad q_e = q^*, \qquad V_e = V^*, \qquad k_e = k^*$

21 $\qquad\qquad r_e = w_e = Cc, \qquad Y_e = A\langle Cc\rangle,$

The equilibrium state is *non-Walrasian*, because of the presence of commercial stocks at the trader and backlog of orders at the producers in equilibrium.

22 Viability analysis

The viability set, the set of admissible inputs and their compatible subsets, is the following.

$$\mathcal{X}^C := \{q, V, k, r \mid q \geq 0, V \geq 0, k \geq 0, 0 \leq r \leq \bar{r}\},$$

$$\mathcal{U}^C := \{q^*, V^*, k^*, c \mid q^* > 0, V^* > 0, q^* > 0, c \geq 0\},$$

$$\mathcal{U}_1^C := \{(q_1^*, V_1^*, k_1^*, c_1) \in \mathcal{U}^C \mid Cc_1 < \bar{r}\},$$

$$\overset{\circ}{\mathcal{X}}_1^C := \{(q_1, V_1, k_1, r_1) \in \overset{\circ}{\mathcal{X}}{}^C \mid (E-A)r_1 \geq 0\}.$$

Since w and Y are not restricted in sign, the trader may withdraw the order not filled by the producer and the user may make return transfer to the trader. $\mathcal{X}^C$ and $\mathcal{U}_1^C$ are non-empty.

The *viability theorem* for Model C sounds like that for the earlier models.

27 Conclusions

The introduction of a separate commercial sector into our family of models necessitated reinterpretation of some processes and variables. Model C still remained very similar to Model SB.

a) The producers who decide on their production level (knowing the unfilled orders of the trader) transfer the total produced quantity to the trader. They, as users of input materials, also decide on the purchases as a function of their input stocks. They and the end users are served from the commercial stock kept by the trader. The trader decides on the placement of orders on the basis of his (commercial) stocks.

b) In spite of this change in the model structure the basic characteristics of the control system remained unchanged as compared with Model SB. (Vegetative control with transactional, quantity, communication; PI type controller; stability; non-Walrasian equilibrium; viability)

c) The primary effect of the introduction of commerce was to split the market of each commodity into two: the wholesale and the retail market. Since the wholesale market is supplied by the producers on the order of the trader and the retail market is supplied from the commercial stock, the first one is a seller's market, the second one a buyer's market, and — taken both together — the market is orderless. The role of the trader is precisely that of absorbing market imperfections.

A second effect of commerce is a saving in information processing and transmission. The commercial sector functions accessorily as an information gathering and condensing agent.

Chapter 13

THE SUPPLY-SIDE PRICE SIGNAL (MODEL P)

1 *The role of the price signal*

The models discussed so far have been non-Walrasian not only in the sense that in the equilibrium state the supply = demand equality does not hold, but I have deviated from the Walrasian paradigm also in disregarding prices, its dominant signalling system. Thus the suspicion may arise that there is a connection between the two omissions; if prices do not affect production and turnover of goods, then of course Walrasian equilibrium cannot be attained. In the Theory of Equilibrium with Rationing *the* cause of the deviation from Walrasian equilibrium is usually imputed to "sticky" prices, by which it is meant that prices do not react, or else do not react with sufficient speed and force to the excess supply or excess demand. In this chapter I present Model P (P for price) in which prices immediately follow the changes in the output stocks and production reacts without delay to the change in prices (more precisely: profitability), but the resulting steady state is still not a Walrasian equilibrium.

In this treatise I refrain, as a rule, from drawing practical lessons from my oversimplified models. At this point, nevertheless, I would like to call attention to something which you may call, if you so wish, a practical conclusion. Non-Walrasian phenomena occur in capitalistic economies (permanent unemployment, unsaleable output stocks, excess productive capacities) as well as in socialist ones (e.g. shortage in consumer goods, materials, manpower, hoarding of input stocks). There are many economists who tend to find the explanation of these phenomena in "false" prices, as represented, for example, by high wages attained by the trade unions, regulation of the prices or other restrictions of the market by the state (in capitalistic economies), or by fixed prices of consumer goods and raw materials, state regulated wages (in socialist economies). And if this diagnosis is right, then the treatment can only be to replace the false price with the "right", i.e. Walrasian equilibrium, price. My studies raise at least some doubts as to both the correctness of the diagnosis and the efficiency of the therapy.

In this chapter, however, I only deal with one simple price setting principle; namely the setting of the price of any product by its producer, which I called supply-side price. Other price setting principles will be dealt with in Part Three.

2 Price and profitability

In contrast with the previous four models where there were only quantity signals (stocks and orders) here I will introduce the prices, represented by the price vector p. Profitability of a product will be measured as *value added per unit product*, i.e. the difference between the price of the product and that of the inputs required for its production:

3
$$g := p - \mathbf{A}'p = (\mathbf{E} - \mathbf{A}')p,$$

where $\mathbf{A}'$ is the transpose of the input coefficient matrix $\mathbf{A}$.

The price p and the indicator g will be used only as intermediate signals, "information used in making decisions on production", and will fail to execute many other functions which in our minds are usually imputed to prices. ("Accounting prices".)

This narrowing down of the role of the price, and the fact that money flows do not appear in Model P, make this exercise weak compared with neoclassical price-signal models. My model reflects very little of how actual price changes affect a real economy, and so I cannot claim it as a rival. In a nutshell and without pretensions to completeness, the following functions of price and money fall outside the scope of the present study of Model P.

a) *Generation, flow and spending of income.* Even the question of whether the buyers have enough money to pay for the purchases cannot be asked in this framework. (Thus viability will not be restricted by budget constraints.)

b) *The incentive role of prices; interest in making profit.* In my model the producers also react to price changes, but the neoclassical theory offers a much richer illustration of how the motivation of the agents is affected by the prices.

c) Prices are known to play an important role in the efficient *allocation of resources*. With the omission of technological choice this aspect has also been excluded from the present study.

Thus the role of Model P is simply that of demonstrating that the introduction of prices in this limited role does not much alter the operating characteristics of a model structurally similar to the previous ones.

4 The real processes

We will begin the description of the real processes of the economy in the usual way:

5
$$\dot{q} = r - Ye - c$$

6
$$\dot{V} = Y - A<r>$$

7 Assumption on the input stock equilibrium

Since the novelty of Model P consists of the introduction of prices into the decision making on the production level, where profitability will play a role, while prices will not directly affect purchases (speculative material purchases are thus omitted), it seems not to be too strong an assumption that the system starts at and remains in equilibrium on the material input side, i.e. we assume for the initial values of V and Y:

8
$$V_0 = V^*$$

9
$$Y_0 = A<r_0>$$

The input stock can be kept on the normal level for all $t > 0$, if material purchases follow the rule

10
$$Y = A<r>.$$

This implies a purchasing behaviour of buying as much material at each instant as is immediately put to use in production. Since the decision on production will depend on the prices, so, indirectly, will the purchasing decisions.

11 *Reformulation of the real processes*

Substituting from **10** to **5** the output stock balance reduces to

12 $$\dot{q} = (E - A)r - c.$$

13 *The control process*

The control of production will consist of three subprocesses:

output stock → price setting → profitability calculation → production level.

PRICE SETTING

14 $$\dot{p} = -2\beta\gamma\dot{q} - \gamma^2(q - q^*).$$

Each price is set by the producer of the commodity in question. The price is raised if the output stock is below the normal level and/or if the stock falls, and conversely. (Recall 3.9 about the observed dependence of prices on output stocks.)

PROFITABILITY CALCULATION

15 $$g = (E - A')p$$

as in **3**.

PRODUCTION DECISION

16 $$\dot{r} = \alpha\dot{g},$$

where $\alpha > 0$ is a new control parameter. The equation expresses that production rate is proportional to the rate of profitability. (Negative α, i.e. decreasing production in response to growing profitability, would be highly unreasonable.)

17 Structural analysis

The structure of decision making on production in Model P is similar to that of Model S, the stock signal model. The source of the similarity is that in both cases the producers keep output stocks and their level provides the principal information on which the decision on production is made. The difference is that in the present model the observation of the output stocks is followed by three subprocesses. The first one, the price setting process (**14**), is formally similar to previous behavioural equations, with the difference that it does not produce a manipulating variable, but an interim control variable, i.e. the price. Each supplier sets the price of his product in an uncoordinated way, relying only on the observation of his own output stock. But the signal generated by this process is then communicated to all the other producers so that they can make the profitability calculations for which all these prices are necessary. The transmission of the price signal can be interpreted in the simplest way (i.e. without the involvement of a superfluous new organization) if we assume that the suppliers transmit the prices they set directly to the interested buyers. Thus the price setting subprocess is *transactionally communicative*. Its difference from the similarly communicative order signal models is that the contents of the signal are no longer a quantity but a price signal.

In the next step (**15**), each producer calculates the profitability of his product for which no information is needed other than the price signals he received, his own price and his technological coefficients. Thus this subprocess is uncoordinated. It is also non-communicative since in the next step (**16**) the production decision is made with the knowledge of his own profitability indicator only.

Finally the decision on transfer (**10**) is made by the buyers corresponding to their own decision on production.

Summing up: Model P represents a *vegetative control* system *with uncoordinated price setting by the supplier* and with *transactional price communication* between the producers.

The simple (single-loop) block diagram of Model P is shown in Figure 10.

The *market structure* of Model P is also similar to that of Model S; this is also an orderly market with the buyers on the short side. (Buyer's market.)

Figure 10

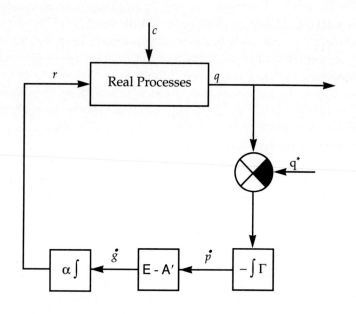

Block diagram of Model P

18 Stability analysis

This model cannot be brought to the reduced form so instrumental in the stability analysis of the previous four models. We must proceed differently and return to Theorem 7.22.

Let us first take the derivative of equation **15** with respect to time, substitute $\dot{g}$ into **16**, and then substitute $\dot{p}$ from **14** into the resulting equation. In this way we obtain:

19 $$\dot{r} = \alpha(E - A')[-2\beta\gamma\dot{q} - \gamma^2(q - q^*)].$$

Let us provisionally put

20
$$\beta \to \frac{\beta}{\sqrt{\alpha}}$$
$$\gamma = \frac{\gamma}{\sqrt{\alpha}},$$

so that α formally disappears from equation **19**, its role to be discussed later. Let us furthermore substitute $\dot{q}$ from **12** into **19** to obtain

$$\dot{r} = (E - A')[-2\beta\gamma(E - A)r + 2\beta\gamma c - \gamma^2(q - q^*)].$$

This equation and equation **12** can then be put into the following state equation form:

21
$$\begin{bmatrix} \dot{q} \\ \dot{r} \end{bmatrix} = \begin{bmatrix} 0 & E - A \\ -\gamma^2(E - A') & -2\beta\gamma(E - A')(E - A) \end{bmatrix} \begin{bmatrix} q \\ r \end{bmatrix} +$$

$$+ \begin{bmatrix} -c \\ (E - A')(\gamma^2 q^* + 2\beta\gamma c) \end{bmatrix}.$$

Let us introduce the notation:

$$D := (E - A')(E - A)$$
$$\mathcal{D} := \{\delta_1, \ldots, \delta_n\}, \text{ the spectrum of } D$$

22 $$H := \begin{bmatrix} 0 & (E - A) \\ -\gamma^2(E - A') & -2\beta\gamma D \end{bmatrix},$$ the system matrix in **21**

$$\mathcal{H} := [\chi_1, \chi_2, \ldots, \chi_{2n}], \text{ the spectrum of } H.$$

23 *Lemma*

a) If $\delta \in \mathcal{D}$, then δ is real and positive, $\delta > 0$.

b) $\chi \in \mathcal{H}$ if and only if it is a root of the characteristic equation

24 $$\chi^2 + 2\beta\gamma\delta\chi + \gamma^2\delta = 0 \quad \text{for some } \delta \in \mathcal{D}.$$

c) For all $\chi \in \mathcal{H}$, Re $\chi < 0$ if and only if $\beta > 0$.

25 *Proof of Lemma* **23**

a) D is seen to be a non-singular *Gram*-matrix of the form M'M which is symmetrical, and hence has only real eigenvalues. It is also known to be positive definite, i.e. $\delta > 0$, $\forall \delta \in \mathcal{D}$.

b) Let us denote by χ an eigenvalue of H and by

$$h = \begin{bmatrix} h_1 \\ h_2 \end{bmatrix}$$

the corresponding eigenvector. Hence we must have $Hh = \chi h$, i.e.

$$\begin{bmatrix} 0 & E - A \\ -\gamma^2(E - A') & -2\beta\gamma D \end{bmatrix} \begin{bmatrix} h_1 \\ h_2 \end{bmatrix} = \chi \begin{bmatrix} h_1 \\ h_2 \end{bmatrix}$$

or row by row:

26 $$(E - A)h_2 = \chi h_1$$

27 $$-\gamma^2(E - A')h_1 - 2\beta\gamma D h_2 = \chi h_2.$$

Multiplying **27** by χ and substituting into it χh_1 from **26** we obtain

SUPPLY-SIDE PRICE SIGNAL

$$(-\gamma^2 - 2\beta\gamma\chi)Dh_2 = \chi^2 h_2.$$

Hence h_2 is an eigenvector of D, and denoting the corresponding eigenvalue of D by δ we have

$$\delta = \frac{\chi^2}{-\gamma^2 - 2\beta\gamma\chi}$$

which yields the characteristic equation **24**.

c) Solving **24** for χ we get for $\delta \in \mathcal{D}$

$$\chi = \gamma[-\beta\delta \pm \sqrt{\beta^2\delta^2 - \delta}\,]$$

Since $\gamma > 0$ and $\delta > 0$ by a) we find that Re $\chi < 0$ if and only if $\beta > 0$.

We have also proved that the elimination of α from **19** has not affected condition c), since the old and the new β have the same sign.

If Re $\chi < 0$ for all $\chi \in \mathcal{H}$ then by virtue of Theorem 7.**22** the system **21** is stable. This proves stability of the original system in q and r. Stability in Y then follows from **10**, in V from $V(t) = V^*$ (a consequence of Assumption 7) and in p and g from **14** and **15**, respectively. Thus we have proved the following theorem.

28 *Stability theorem for Model P*

If the input coefficient matrix **A** satisfies Assumption 8.7 and the input stock V starts in equilibrium in the sense of Assumption **7**, then the Model P consisting of the controlled subsystem **5-6** and the controller **14-16** is stable if and only if $\beta > 0$.

29 *The constant-input steady state*

Putting $c(t) = c$, $\forall t \geq 0$ and all the derivatives equal to zero, we get for the stable system satisfying the conditions of **28** the steady-state:

30 $\qquad\qquad q_e = q^*, \qquad V_e = V^*$

31 $\qquad\qquad r_e = Cc, \qquad Y_e = A\langle Cc\rangle$

The steady prices can be calculated by integrating **16** from $t = 0$ to $t = +\infty$ to get

$$r_e - r_0 = \alpha(g_e - g_0) = \alpha(E - A')(p_e - p_0)$$

from which we express p_e:

32
$$p_e = p_0 + \frac{1}{\alpha} C'(Cc - r_0).$$

It can be seen that the steady prices (in contrast with the other variables) depend on the initial value of p.

The equilibrium (including the equilibrium prices) is non-Walrasian because q_e is positive.

33 *Viability analysis*

With respect to Model P, where V is kept constant and where Y is non-negative if r is **(10)**, the sets we have to define are simpler:

34 $$\mathcal{X}^P := \{q, r \mid q \geq 0,\ 0 \leq r \leq \bar{r}\}$$

35 $$\mathcal{U}^P := \{q^*, c \mid q^* > 0,\ c \geq 0\}$$

36 $$\mathcal{U}_1^P := \{(q_1^*, c_1) \in \mathcal{U}^P \mid Cc_1 < \bar{r}\}$$

37 $$\mathcal{X}_1^P := \{(q_1, r_1) \in \overset{\circ}{\mathcal{X}}{}^P \mid (E-A)r_1 \geq 0\}.$$

The *viability theorem* for Model P can again be written out on the pattern of the previous models.

38 *The non-negativity of the prices*

We have not stipulated non-negative prices as conditions of viability. This can be justified from a purely logical point of view. Our prices are accounting prices and we can make calculations with a price vector having some negative components equally well as with all positive prices.

But this is formal argumentation and I wish to eliminate negative prices so as to comply with economic interpretability. To this end I will draw on the parameter α, which has been utterly neglected so far.

Let us assume that the initial prices are positive: $p_0 > 0$. First integrate equation **16** from 0 to t and substitute g in it according to **15**.

$$r(t) - r_0 = \alpha(E - A')(p(t) - p_0),$$

or equivalently

$$p(t) = p_0 + \frac{1}{\alpha} C'(r(t) - r_0),$$

which holds for all $t \geq 0$. But $C' \geq 0$, and $r(t) \geq 0$ (viability), hence

39
$$p(t) \geq p_0 - \frac{1}{\alpha} C' r_0.$$

Now if we choose α as large as

40
$$\alpha > \|<p_0>^{-1} C' r_0\|,$$

then

$$\alpha e > <p_0>^{-1} C' r_0$$

$$\alpha p_0 > C' r_0.$$

Thus the r.h.s. of **39** is positive, and we *conclude that the prices will be positive* all the time if α is as large as to satisfy **40**.

41 *Conclusions*

Model P was similar to Model S but used prices set by the suppliers and profitability indicators calculated from these prices in the decisions on production. Furthermore it was assumed for sake of simplicity that the input stock is kept at normal level and the material purchases are in equilibrium with the production level throughout. We fitted a three-step controller to this model with the following characteristics:

a) The controller is vegetative with transactional (price) communication. The price of each commodity is set by the supplier on the basis of his output stock and is then communicated directly to all buyers. This was done by a PI transfer element. The price setting is followed by calculating the value added per unit product (profitability) effected by the producer. This was performed by a P element as was the next, where the production level was decided by the producer on the basis of the profitability indicator. Thus the whole control process is of a PI character.

b) The stability of the model has been achieved for all $\beta > 0$ in contrast with the other models where β was subject to a lower bound depending on the input coefficient matrix **A**.

c) The constant-input steady state was again non-Walrasian in spite of the fact that prices were adjusted immediately to excess demand (represented by the rate of change of output stocks and its deviation from the norms). The price vector also approached a non-Walrasian steady value which, however, depends on the initial price vector.

d) Viability of the system can be established in the sense that stocks and production remain non-negative and the latter does not exceed the productive capacity. Moreover the non-negativity of the prices can also be achieved for an appropriately large value of a third control parameter.

Chapter 14

SUMMARY OF PART TWO

Having arrived at the end of Part Two I will try to summarize the lesson which can be learned from the collation of the five models we have studied.

1 *Why precisely were these models analysed?*

The reader who has waded through the models discussed so far might well draw the conclusion that he has met individual control structures selected at random. This series may seem to be continuable indefinitely, without the emergence of a systematizing principle which would point the way to the construction of controllers having prescribed properties.

This feeling of lack is justified; the presented models have been piece produced. Behind them is another invisible series of many times five discarded models which either met their death in the waste basket after a cumbersome analysis, or died even earlier in the head of the researcher. It might not have been without point to show what kind of models were judged not worth presenting and why. This might have broken the above dull repetition of similar conditions and the false impression that any "reasonable" model performs well. But I was afraid of overtaxing the reader's patience. Rather, I present a list of the points of view which served as selection criteria.

a) There were a few models, pleasant in mathematical formulation and not hard to analyse, which were discarded for their rather artificial economic interpretation. Some of them implied processes which cannot be carried out in the real world or which meet with a logical hitch.

b) Other models were faultless both from the economic and the mathematical points of view but did not offer more lessons than other simpler models. (This was, for instance, the reason for discarding model variants in which the purchasing behavior, stocks and consumption of the end user were portrayed in more detail.)

c) Since my purpose in Part Two was to prove the viability of the simplest control structures, such as do not require coordination of the agents' decisions, some models which violated this criterion were left out (e.g. one in which the price is set by suppliers and buyers jointly).

d) Nonviable systems were sorted out at once. (Viability in terms of non-negativity and productive capacity constraints.) Their presentation would have yielded the benefit of showing the blind alleys. Nonviability can be a consequence of instability. Such systems might be stabilizable by other kind of controllers (e.g. non-linear ones) which we have not introduced for sake of mathematical simplicity. Nonetheless, there are simple linear controllers which can be stabilized but where the system still cannot be made viable. Let me give an example: the use of a proportional (P) controller instead of the proportional plus integrating (PI) one we applied. Such a system can be brought to the form below (Cf. 6.**40-41**, $R_2 = 0$ is assumed for the sake of simplicity):

$$\dot{y} = P_1 y + P_2 v + R_1 z$$
$$v = P_3(y - y^*).$$

This system represents uncoordinated control whenever P_3 is diagonal. It can be stabilized if the matrix $P = P_1 + P_2 P_3$ has eigenvalues with negative real parts for some diagonal P_3. But it is easy to see that if a component of y can (and why should it not?) take values both above and below the norm, then the corresponding component of v cannot satisfy a non-negativity constraint.

e) Finally — alas — there were cases where I could not tell if the models were stable or under what conditions they would be stabilisable. Here I met the most regrettable constraint, that of my knowledge or inventiveness. In order still to be able to say everything what I thought important, I was in these cases forced to compromise. One fortunate result of such a search for a way out was the "trader" of Model C, another really painful compromise was to assume constant input stocks in Model P.

2 *Common starting points of the models*

The five models we studied in Chapters **8-13** set out from the following common properties:

a) The *real processes* of the models were alike: an open Leontief-economy with a constant input coefficient matrix. There are as many producers as products. (Plus a trader in Model C.) The *end use* is an exogenous function of time.

b) *Scarcity of resources* was represented only indirectly by productive capacity constraints, which also restricted end use indirectly.

c) *Monetary processes*, fiduciary goods were neglected, budget constraints did not restrict real activities.

d) Both the real processes and the control processes evolved in *continuous time* and were represented by linear differential equations. There was no time lag in these equations: production, transfer, information processing and decision making were instantaneous.

e) To the common real sphere five controllers were fitted, differing in their information and decision structure, but consonant in:
— consisting of proportional and integrating transfer elements (PI character)
— complying to constant norms as command signals (stock norms and backlog of order norms)
— applying uniform control parameters in the control of different processes.

3 *Tabular survey of the deviations among the models*

The major deviations among the five models can be read off from Table 4.

Table 4

Sign of the model Main characteristic	(S) Stock signal	(B) Order signal	(SB) Stock and order signal	(C) Commercial sector	(P) Price signal
Controlled variables and command signals					
Input stock and its norm	+	+	+	+	−
Output (or commercial) stock and its norm	+	−	+	+	+
Interim control signals					
Order signal, backlog and its norm	−	+	+	+	−
Price and value added	−	−	−	−	+
Who decides?					
The producer (P) or the trader (T) as buyer (B) or as supplier (S)					
Transfers	PB	PS	PS	PS, PB	PB
Order placements	−	PB	PB	T	−
The structure of the control circuit					
Single-loop (S), multi-loop (M)	S	M	M	M	S (Three sub-processes)
Stability					
(zero or positive lower bound on β)	+	+ if stabilizable	+	+	0
Viability set					
Is return transfer (RT) or withdrawal of orders (WO) permitted?	RT	WO	WO	RT, WO	RT
Degree of centralization					
Vegetative (V), non-communicative (N), with quantity (Q), or price (P) communication	VN	VQ	VQ	VQ	VP
Market structure					
Buyer's (B) seller's (S), or orderless (OL) market	B	S	OL	OL Wholesale: S Retail: B	B

4 What was I able to demonstrate?

The analysis of the five models demonstrated that without exaggerated assumptions (the really strong assumptions were made by excluding important aspects from the study and in the manner of building up the models) the following lessons can be drawn:

a) The systems were proved to be *stabilizable* by choosing the damping exponent β sufficiently large. The lower bound on β depends on the technology matrix only (save for Model B where stabilizability was not guaranteed for all such matrices).

b) Under some compatibility assumptions concerning the viability set, the admissible input set and the initial state, the systems had *viable solutions*. This meant that non-negativity and capacity constraints could also be satisfied for variable end use if it remained in a neighbourhood of the initial net output.

c) Since all five models were *vegetative*, it was proved that *coordination* and *control organizations are not needed* to perform the functions included in the model for the economy to survive. But—except for the simplest case of Model S—there is a need for (transactional) communication among the agents; in our case orders or prices were communicated.

d) The same real economy—on the given level of abstraction—could have been successfully controlled with *controllers of different information and decision structures*. This proposition casts doubts on the belief that to each real economy there corresponds one and only one controller (economic mechanism) by which it can be appropriately regulated.

e) The "markets" which we studied were generically *non-Walrasian* in the sense that:
 — they applied a non-Walrasian signalling system, quantity signals and non-Walrasian prices;
 — their steady (= equilibrium) state for constant end use was non-Walrasian, because useless output and/or input stocks and/or backlog of orders remained in existence forever;
 — the markets were orderly in some models and orderless (willing buyers and sellers did not necessarily transact) in others; there

was no essential difference in the functioning of these systems.

(I have put *market* in quotation marks above in order to indicate that I have not dealt here with the atomistic or competitive markets traditionally studied within the Walrasian paradigm.)

5 And in what did I fail?

The points with which I am dissatisfied in connection with the analysis of the models are the viability theorems.

a) The existence of viable solutions cannot always be decided. The conditions under which viability has been demonstrated are sufficient but not necessary.

b) Even if viability is established by my method, the domain of viable solutions is much more restricted than necessary. The estimates are too rough.

c) I was not able to establish how this domain could be enlarged by the choice of the control parameters. The parameter γ, which has not played a substantial role in the analysis, may exert such an effect. This is why I have not discarded it (putting it equal to one) in the hope that one day I myself or someone else can use it for this purpose.

In all events, I encourage the interested reader to try to apply new ideas to those problems which I was not able to solve satisfactorily.

6 The controversy about "free market forces" vs "state intervention"

I feel an urge to clear up a possible misunderstanding of what I have written. Am I suggesting that in real life economy the coordination of economic activities is superfluous (or even harmful), and that the economy should be left alone to follow its own path, or—as is often said—the best economy is the one where market forces freely exert their effect? No, I am definitely not suggesting that. I am not against the role which the state and other social organizations can play in controlling the economic processes.

In my opinion — but this is merely an opinion supported only fragmentarily by scholarly arguments — there are three kinds of processes in each economy:

(i) processes in which state control is superfluous and often harmful;

(ii) processes which — to some extent — can be subjected to intervention by the state, this intervention being usually beneficial according to some criteria but incurring losses according to others. The balancing of such incommensurable effects is the "art of economic policy making";

(iii) processes which, without control by the state, could disrupt the economy.

I am not in a position to draw the boundaries between these three kinds of processes — and the adverb "yet" has been omitted from this sentence deliberately.

What then is the purpose of demonstrating viability of some economic processes (like production and turnover of commodities) without coordination? It is that if economics as a science will one day be able to draw the boundaries where state intervention should begin, this boundary must be sought in processes which are omitted from the above — vegetative — models. Let me name just a few of them:

— allocation of primary resources,
— technical progress,
— international economic relations,
— the investment process,
— monetary processes,
— peculiarities of the labour market

and others. Which ones should be coordinated and which others not; this is what ought to be made clearer by further research into the theory of economic mechanisms.

BIBLIOGRAPHICAL NOTES TO PART TWO

A paper by KORNAI and MARTOS (1973, and NPC, Chapter 2) analysed a model which can be considered as the archetype of the models discussed in Part Two. It already contained the open Leontief-economy as the representation of the real processes (with exogenously changing technology), a PI type controller (without assuming uniform parameters), input and output stock norms. Many more ideas explained more fully in the present book already appeared there in an embryonic form.

As Professor *A. Mátyás* pointed out to me, many of the ideas which led to the construction of these types of models can be traced back to the Fixprice Method of HICKS (1965). His term "desired stock" can be identified with what I call "normal stock" and he defines as "stock equilibrium" the state where actual stock equals the desired stock, which is also the case with my "non-Walrasian steady state". But the most interesting relationship is that between Hicks' "stock adjustment principle" and my PI controller. Let me illustrate this first by a quotation from HICKS (1974, p.15):

> It has become conventional (since the time when it was recognized that the level of stocks must not be neglected) to suppose that it is regulated by what is called the *stock adjustment principle*. Producers (and traders) are supposed to have some desired level of stocks; so their demand for replacements is governed, in the first place, by the rate of sales which they expect, and in the second place by the difference between actual stock and desired stock. It is not supposed that they will seek to reach their desired stock instantaneously; they will plan to work up their stocks, or to work them down, over an appreciable period.

In this original form the stock adjustment principle refers to two terms in production control: expected sales and stock deficiency. I will return to the first term, the second being similar to the proportional term in my production control equation. He also considers taking an additional differentiating term (HICKS, 1965, p. 99) into account, and argues that this is a very natural behaviour, but concludes by rejecting the study of such a model because the stability analysis would be hard to carry out. (He is right concerning his model where there is a production lag and a different decision lag.) I myself opted for an integrating term, which is no less reasonable, and here we have parted company.

Returning now to the expected sales, it is worth mentioning that something similar (actual sales) appeared on the right hand side of the production control equation in the archetype model mentioned above. The present Model S in *Chapter 9* differs from the archetype model in the elimination of this term (the derivative of $Ye + c$) from the counterpart of equation 9.4 and a similar omission of the derivative of the material input $A<r>$ from that of 9.5. Model S proved that such terms are not needed; neither actual nor expected sales have to be taken into account.

The motivation behind this omission was, however, the following. In contrast with the intention and claim of the authors of the Kornai-Martos model, it could not be interpreted as representing a vegetative (uncoordinated) system. At the time I failed to perceive, and discovered only recently, that the additional terms imply simultaneous decision making on production and transfers by all producers, that these decisions depend on each other; thus they cannot be made without coordination by a central control organization. The elimination of the culprit terms cut this interdependence of the producers' decisions, but made the analysis of the resulting Model S more difficult and less spectacular.

Model B of *Chapter 10* is a cross-breeding of Model S and the order signal model of KORNAI-SIMONOVITS (NPC, Chapter 12). The model of Kornai and Simonovits was another descendant of the "archetype" model discussed above, but crucially different in that they worked with discrete time, proportional controller and variable command signal. A good discussion of the production to order *vs* production to stock can be found in the book by BELSLEY (1969).

Model BS of *Chapter 11* and Model C of *Chapter 12* are both new combinations of the ideas which led to Models B and S.

In the construction of Model P of *Chapter 13* I borrowed the idea of using the "value added" indicator as a control signal from BRÓDY (NPC, Chapter 6), another follow up to the Kornai - Martos paper. Bródy himself refers to GOODWIN (1953) as the originator of the idea. Bródy's stability proof also gave hints for the stability analysis of the more complicated Model P.

Other models related to those discussed in Part Two can be found in Non-Price Control, especially in Chapters 5, 6, 10-13 by *Dancs, Hunyadi* and *Sivák, Bródy, Kornai* and *Simonovits, Kapitány, Simonovits*.

PART THREE

EQUIVALENT CONTROLLERS, PARTIAL COORDINATION

In the third part of the treatise I deal with two problems. The first one is this: when can we consider two controllers fitted to the same real processes to be equivalent in their operation? I deal with this problem in terms of Laplace transforms on a moderate level of abstraction. The use of Laplace transforms also facilitates the study of the second problem: how can we construct price signal models which are equivalent with another model using quantity (stock) signals only. In this way I can obtain several systems of different information and decision structures, which, however, have this much in common, that certain decisions must be coordinated either in an interactive way or rather by specialized control organizations. Thus the emphasis is here shifted from stability and viability analysis to the study of these organizational aspects of coordination.

Chapter 15

THE LAPLACE TRANSFORMATION

1 *On the Laplace-transform: economic interpretation*

We may begin by acknowledging that the Laplace-transformation is not one of the techniques frequently used by economists. Therefore, by way of introduction, we will attempt to interpret the essence of the Laplace-transform in economic terms — despite the fact that this interpretation is not quite correct and cannot even be employed in that part of the book where the transform is used.

Let $x(t)$ be a scalar or vector-valued function of time t defined in the domain $t \geq 0$. To facilitate an economic interpretation let us assume, for example, that $x(t)$ denotes an outlay to be effected at some date t. Let $\mathbf{s}$ be the discount rate, then the present discounted value of the process $x(t)$ is given by the integral

$$\int_0^\infty x(t)\varepsilon^{-st}\mathrm{d}t.$$

Obviously, knowledge of this present value is not sufficient for the reconstruction of the original process; processes with different time distributions may yield the same present value. But could the function $x(t)$ be reconstructed if we knew its present value not only for a single discount rate, but as a function of $\mathbf{s}$?

A positive answer may be given for any function for which the above integral exists. In practice, this requires only that the function $x(t)$ should not tend towards infinity more rapidly than an arbitrary exponential function. If the function $x(t)$ is continuous, the convergence of the integral requires no other condition. If we now consider the above present value as a function of the variable $\mathbf{s}$, and denote it by $\mathbf{x(s)}$, then the function

2
$$x(s) = \int_0^\infty x(t)\varepsilon^{-st}dt$$

is called the Laplace-transform of the function $x(t)$. (As here, the Laplace-transform of a variable is denoted by bold-face type of the same letter. In most cases this allows us to omit the arguments t and s without risking confusion.) In what follows the existence of the Laplace-transform is assumed for each of our functions, without each being individually examined.

Table 5

Laplace Transformation: examples			
Function number	$x(t)$, $t \geq 0$	$x(s)$	Domain of validity
1	1	$\dfrac{1}{s}$	Re $s > 0$
2	t	$\dfrac{1}{s^2}$	Re $s > 0$
3	t^n	$\dfrac{n!}{s^{n+1}}$	Re $s > 0$, n integer
4	$\varepsilon^{-\alpha t}$	$\dfrac{1}{s+\alpha}$	Re $s > -\alpha$, α real
5	$\sin \omega t$	$\dfrac{\omega}{s^2 + \omega^2}$	Re $s > 0$, ω real
6	$\cos \omega t$	$\dfrac{s}{s^2 + \omega^2}$	Re $s > 0$, ω real

Now, for all practically relevant cases it turns out that there is a one-to-one correspondence between the original function of time and its Laplace-transform; the one can be produced from the other. While the formation of the Laplace-transform involves using the above integral, the inverse transformation is produced by the following complicated formula which cannot be used for practical purposes:

3
$$x(t) = \frac{1}{2\pi \iota} \int_{\sigma-\iota\cdot\infty}^{\sigma+\iota\cdot\infty} x(s) \varepsilon^{ts} ds,$$

where the real constant σ must be chosen so as to exceed the largest real part of the poles of the function $x(s)$.

Instead of using formulae **2** or **3**, the Laplace-transform and — even more so — the inverse transform are simply taken from function tables. We also present here a small example of such a table (Table 5).

With the inverse transformation, it is clear that there is no longer any possibility of an economic interpretation. So far we have spoken of the variable s as if it denoted a real value. But in the theory of the Laplace-transform s is a complex number. We cannot, however, give an economic interpretation to a complex valued discount rate.

To denote the Laplace-transform we shall sometimes use the symbol $\mathscr{L}[\cdot]$, and to denote the inverse transform by the symbol $\mathscr{L}^{-1}[\cdot]$. Thus, in terms of this notation:

$$\mathscr{L}[x(t)] = x(s)$$

$$\mathscr{L}^{-1}[x(s)] = x(t).$$

4 Some properties of the Laplace-transform

We summarize some properties of the Laplace-transform in Table 6.

The linearity properties **1** and **2**, according to which sums may be transformed term by term, (leaving the time-independent coefficients unchanged) are particularly important for us. The derivative properties **5** and **6**, showing that the Laplace-transformation turns differential equations into ordinary (algebraic) equations in terms of s, are no less important. With items **5** and **6** the subscript $_0$ refers to the initial ($t = 0$) value of x (and its derivatives) in the event of their being continuous from the left at point 0. In the event of discontinuity (switching on), the left-hand limit (t→ −0) is to be taken.

Table 6

Number of property	x(t), t≥0	x(s)	Name of property
	LAPLACE TRANSFORMATION: RULES		
1	$\alpha_1 x_1 + \alpha_2 x_2$ (α_1, α_2 real)	$\alpha_1 \mathbf{x}_1 + \alpha_2 \mathbf{x}_2$	Linearity
2	Ax (A is a constant matrix)	$A\mathbf{x}$	Linearity
3	$x(\dfrac{t}{\alpha})$ (α is real)	$\alpha \mathbf{x}(\alpha s)$	Similarity
4	$\varepsilon^{-\alpha t} x$ (α is real)	$\mathbf{x}(s+\alpha)$	Damping
5	$\dot{x}$	$s\mathbf{x} - \mathbf{x}_0$	Derivative
6	$\dfrac{d^n x}{dt^n}$	$s^n \mathbf{x} - s^{n-1}\mathbf{x}_0 - \sum_{k=2}^{n} s^{n-k}\left(\dfrac{d^{k-1}x}{dt^{k-1}}\right)_0$	Derivative
7	$\int_0^t x\,d\tau$	$\dfrac{\mathbf{x}}{s}$	Integral
8	$x(t+\alpha)$ ($\alpha \geq 0$)	$\varepsilon^{\alpha s} \mathbf{x}$	Shift in time-domain

5 Laplace-transform of the transfer element

The Laplace-transform of a time-invariant linear transfer element is written as follows:

$$\mathbf{y} = \mathbf{F}\mathbf{v}, \qquad (6)$$

where **v** is the transform of the input vector, **y** that of the output vector, and the matrix **F**, which is now not constant, but itself a function of **s**, is called the *transfer matrix*.

7 Transition from the state equation form to the $\mathcal{L}$-transform

Let us first consider the state-equation form of a transfer element (Cf. 6.8-9) with constant matrices **P**, **R** and **S**:

8
$$\dot{x} = Px + Ru$$

9
$$y = Sx.$$

By applying properties **2** and **5** of Table 6 the Laplace-transforms of the two equations will be:

10
$$sx - x_0 = Px + Ru$$

11
$$y = Sx.$$

From **10** we obtain:

$$x = (sE - P)^{-1}(Ru + x_0),$$

and by substituting this into **11**

12
$$y = S(sE - P)^{-1}(Ru + x_0).$$

Introducing the notation

13
$$v := \begin{bmatrix} u \\ x_0 \end{bmatrix}, \quad F := \begin{bmatrix} S(sE - P)^{-1}R & 0 \\ 0 & S(sE - P)^{-1} \end{bmatrix}$$

we immediately obtain the equation of the element in the form **6**.

As is apparent from this transformation, when applying Laplace-transforms it is expedient to include x_0, the initial values of the state-variables, in the vector of input signals. (This is only logical; the initial state of the element comes from the past, that is, from outside.)

14 $\mathcal{L}$-transform of a control circuit

The same procedure can be applied to the state-equation form of a control circuit 6.**40-41**, e.g. if considered as a transfer element according to 6.43. But in order to keep the controlled subsystem and the controller apart, we will produce the Laplace-transform of the control circuit in the following form

15 $\quad\quad y = Gy_0 + Hv + Iz \quad\quad$ (The controlled subsystem)

16 $\quad\quad v = Iv_0 + M(y - y^*) + Nz \quad\quad$ (The controller).

The comparison with 6.40-41 gives:

17 $\quad\quad G := (sE - P_1)^{-1}, \quad\quad H := GP_2, \quad\quad I := GR_1$

18 $\quad\quad L := (sS - P_4)^{-1}, \quad\quad M := LP_3, \quad\quad N := LR_2.$

The form **15-18** of the control circuit *will not be used* in the sequel. It has been written down, however, because if someone first puts any model of Part Two in state-equation form and then applies the Laplace transformation, a representation in this form will be obtained. But it is much simpler to apply the $\mathcal{L}$-transformation directly to the original equations (which are not in state equation form). By this trick and others we will be able to produce a simpler $\mathcal{L}$-transform representation, still rich enough to convey the message I am trying to unfold.

19 *The simplified $\mathcal{L}$-transform of a control circuit*

Although the equivalence analysis which follows could be affected also by the use of the formulation **15-18**, I suggest here a simpler form in which I will take into account some properties of the models under study, e.g., that the external effect (end use) does not enter the controller and that the initial values can be eliminated by a coordinate transformation which does not affect the results. However, all this will be demonstrated in the course of the analysis of the individual models, and we may now rest satisfied with making ourselves familiar with the simplified $\mathcal{L}$-transform of a control circuit which reads:

20 $\quad\quad\quad y = Hv + Iz \quad\quad$ (The controlled subsystem)

21 $\quad\quad\quad v = M(y - y^*) \quad\quad$ (The controller)

The control system $\mathcal{S}$ represented by equations **20-21** will be denoted by the following shorthand:

22 $$\mathcal{S} := \{H, I, M, y^*\}.$$

23 *Another form: command transfer and disturbance transfer*

By eliminating **v** from equations **20-21** we obtain

$$y = HMy - HMy^* + Iz,$$

or, assuming nonsingularity of $(E - HM)$ for almost every complex value of **s**:

24 $$y = -(E - HM)^{-1}HMy^* + (E - HM)^{-1}Iz.$$

Let us introduce the notation:

25 $$Y^* := -(E - HM)^{-1}HM = E - (E - HM)^{-1}$$

26 $$Z := (E - HM)^{-1}I.$$

This way **24** turns into the simple form:

27 $$y = Y^*y^* + Zz,$$

where Y^* is called *command transfer matrix* and **Z** is called *disturbance transfer matrix* for obvious reasons.

Chapter 16

EQUIVALENCE OF CONTROLLERS

1 The concept of equivalence

A relation between two elements of a non-empty set $\mathcal{H}$ is called *equivalence relation* (or briefly: equivalence) and marked by $\approx$, if it has the following three properties:

a) $\forall h \in \mathcal{H}$: $h \approx h$, (reflexivity) i.e. each element is equivalent to itself.

b) $\left. \begin{array}{l} h_1, h_2 \in \mathcal{H} \\ h_1 \approx h_2 \end{array} \right\} \Rightarrow h_2 \approx h_1$, (symmetry)

i.e. if an element is equivalent to another, then the other is also equivalent to the first.

c) $\left. \begin{array}{l} h_1, h_2, h_3 \in \mathcal{H} \\ h_1 \approx h_2, h_2 \approx h_3 \end{array} \right\} \Rightarrow h_1 \approx h_3$, (transitivity)

i.e. if two elements are both equivalent to a third one, then they are equivalent to each other.

It is known from the theory of equivalence relations that each equivalence relation partitions the basic set $\mathcal{H}$ into subsets (equivalence classes) $\mathcal{H}_\alpha$, $\alpha \in \mathcal{I}$, (where $\mathcal{I}$ is a finite, or infinite set of indices) such that:

(i) $\bigcup_{\alpha \in \mathcal{I}} \mathcal{H}_\alpha = \mathcal{H}$, i.e. the partitioning is complete,

(ii) $\alpha \neq \beta \Rightarrow \mathcal{H}_\alpha \cap \mathcal{H}_\beta = \emptyset$, i.e. the equivalence classes are disjoint,

(iii) $h_1 \approx h_2 \Leftrightarrow \exists \alpha \in \mathcal{I}$, such that $h_1 \in \mathcal{H}_\alpha$, $h_2 \in \mathcal{H}_\alpha$, i.e. equivalent elements belong to the same equivalence class.

Thus in order to specify an equivalence relation two steps are to be

taken:
— specifying the basic set and
— specifying the rule by which the equivalence classes are formed.

2 Equivalence of controllers: the basic set

As in Part Two the subject of my study remains: how we can fit different controllers, which satisfy certain requirements, to the same real sphere. The five models treated in Part Two were similar in many respects, but the question whether some of them "operate in the same way" has not been raised. (I do not ask this question even now. In the sense of the concept of operating equivalence to be introduced below they are either not comparable, or if they are comparable they belong to different equivalence classes.) If I want now to look for equivalent controllers for a real process, I must first define the set of systems (the basic set) within which the study is to be carried out.

3 Definition of the basic set

The set $\mathcal{H}$ of the control systems to be considered is defined by the following two properties:

a) The system can be represented in the simplified $\mathcal{L}$-transform of a control circuit 15.**20-21** in such a way that 15.**20** represents the real processes and 15.**21** the control processes of the system.

b) For each system $\mathcal{S} \in \mathcal{H}$ the real processes (i.e. the pair of matrices $\mathbf{H}$ and $\mathbf{I}$) are identical: i.e., if $\mathcal{S}_1 = \{\mathbf{H}_1, \mathbf{I}_1, \mathbf{M}_1, \mathbf{y}_1^*\} \in \mathcal{H}$ and $\mathcal{S}_2 = \{\mathbf{H}_2, \mathbf{I}_2, \mathbf{M}_2, \mathbf{y}_2^*\} \in \mathcal{H}$ then $\mathbf{H}_1 = \mathbf{H}_2$ and $\mathbf{I}_1 = \mathbf{I}_2$.

4 Interpretation of Definition 3

The part b) of Definition **3** expresses only our insistence on dealing with controllers fitted to identical real spheres.

The implications of a), however, require further explanation. There is no doubt that if we want to analyse systems equivalence with mathematical tools we have to represent all the systems in a standard mathematical form. The simplified $\mathcal{L}$-transform has been specified with precisely this purpose in mind, that such an analysis could be carried out in a relatively simple way. The crux of the matter is that originally 15.**20** is the equation of the controlled subsystem and 15.**21** of the controller, and this bisection need not agree with the real process vs control process distinction in economic terms. (Let us refer back to 5.7.) This is the reason for setting a stipulation which unifies this double classification. I am not claiming that all linear time-invariant control circuits can be represented in the form 15.**20-21** — obviously not — neither that even if this were possible, the equations could be classified according to assumption a), not to speak of uniqueness of the representation, which obviously does not hold. When all is said and done, I have sacrificed generality for the sake of easier digestibility.

(The interested reader may try his muscle by defining $\mathcal{H}$ on the basis of the more general formulation 15.**15-16** (instead of 15.**20-21**) and then carrying out the subsequent equivalence analysis of the present chapter. He would discover that the same train of thought could be followed but that the formulae would become much more complicated.)

5 *Definition of (operationally) equivalent controllers*

Two systems $\mathcal{S}_1 \in \mathcal{H}$ and $\mathcal{S}_2 \in \mathcal{H}$ are said to be provided with (operationally) equivalent controllers, if for any external effect $\mathbf{z}_1$ and $\mathbf{z}_2$ acting on $\mathcal{S}_1$ and $\mathcal{S}_2$, respectively

6 $$\mathbf{z}_1 = \mathbf{z}_2 (\forall s) \Rightarrow \mathbf{y}_1 = \mathbf{y}_2 (\forall s),$$

where $\mathbf{y}_1$, $\mathbf{y}_2$ is the $\mathcal{L}$-transform of the controlled variables of $\mathcal{S}_1$ and $\mathcal{S}_2$, respectively. That is, an identical path of the external effects evokes an identical path of the controlled variables.

7 *Comments on Definition 5*

The elements of a set can be divided into equivalence classes according to a choice of criteria (possibly joint occurrence of several criteria).

For instance, the population of London can be classified by sex, age, body weight or the time spent in prison. The actual choice of the criterion will obviously depend on the purpose of the study. My purpose has been to answer the following question (asked by *M. Tardos* many years ago):

> Can two systems, one reacting only to quantity signals, the other also to price signals, operate the same way?

The meaning of the phrase "operating the same way" is still ambiguous and might be formalized in several ways. My choice "response to identical external effects should be the same" is only one of the possibilities, in fact a quite strict criterion. If we were guiding a rocket to a target, we would be satisfied with the stipulation that the end point of the paths should coincide. Translating this idea into our own case, we could speak about "asymptotically equivalent" controllers, if done with the equality of the steady states. But as I argued earlier we are at least as interested in the early part of the path as in the later part. In case of asymptotic equivalence, viable and nonviable systems would qualify as equivalent ones.

From a practical point of view, not only identical paths, but neighbouring paths would of course be equally admissible. But in this case an (intransitive) similarity relation, rather than an equivalence relation, ought to be constructed. This procedure would yield less transparent results than the one I opted for.

8 Further comments

a) In the definition of $\mathcal{H}$ nothing has been postulated concerning stability or viability of the systems in question, therefore Definition **5** does not depend on these properties either. But since identical paths are stipulated, a stable control cannot be equivalent to an unstable one, nor a viable with an nonviable one.

b) We have not explicitly assumed anything about the initial values, although it is obvious that different initial values exclude identical paths. The elimination of the initial values from the picture will be carried out by coordinate transformation (translocation of the initial state to the origin).

c) We have not stipulated that the command signals of equivalent systems must be identical. In some cases this will result as a consequence of the equivalence conditions.

9 *Theorem on the equivalence conditions*

Let us consider the two systems

$$\mathcal{S}_1 := \{H, I, M_1, y_1^*\} \in \mathcal{H}$$
$$\mathcal{S}_2 := \{H, I, M_2, y_2^*\} \in \mathcal{H}$$

and assume that $(E - H\,M_1)$ and $(E - H\,M_2)$ are both nonsingular for almost every (complex) value of **s**. The controllers of $\mathcal{S}_1$ and $\mathcal{S}_2$ are equivalent if and only if:

10 $$(E - H\,M_1)^{-1} I = (E - H\,M_2)^{-1} I$$

and

11 $$(E - H\,M_1)^{-1} H\,M_1 y_1^* = (E - H\,M_2)^{-1} H\,M y_2^*.$$

12 *Remark on Theorem 9*

Conditions 10-11 are valid whether any one of the matrices H, I, M_1 and M_2 is square or not. We have never stipulated that any two of the vectors **y**, **v** and **z** be of the same dimension.

13 *Proof of Theorem 9*

Apply implication 6 to systems $\mathcal{S}_1$ and $\mathcal{S}_2$ in form 15.27 to get

$$Z_1 = Z_2$$
$$Y_1^* y_1^* = Y_2^* y_2^*$$

as sufficient and necessary equivalence conditions. Substituting here according to 15.25 and 15.26 we get **10** and **11**, respectively.

14 *Corollary to Theorem* **9**: *special cases*

a) If **I** is of full column rank, then **10** and **11** reduce to

15 $$\mathbf{H\,M}_1 = \mathbf{H\,M}_2$$

16 $$\mathbf{H\,M}_1\mathbf{y}_1^* = \mathbf{H\,M}_2\mathbf{y}_2^*.$$

b) If **I** is of full column rank and **H** is of full row rank, then **10** and **11** reduce to

17 $$\mathbf{M}_1 = \mathbf{M}_2$$

18 $$\mathbf{M}_1\mathbf{y}_1^* = \mathbf{M}_2\mathbf{y}_2^*.$$

19 *Application to Models S and P*

Since we are up to compare models with quantity and price signals, it is reasonable to try our method first on models which have already been studied. For a price-signal model the only candidate is Model P of Chapter 13, while for a quantity signal we choose the one which most resembles Model S of Chapter 9. However, they are still not comparable because with the real sphere of Model P we additionally have assumed input stock equilibrium: $Y = \mathbf{A}\langle r\rangle$ (see 13.**10**) which must now also be extended to Model S. The Model S thus *modified* will be called Model SM.

We are going to show that the stock-signal controller of Model SM and the price-signal controller of Model P are not equivalent. To this end, we must first produce both in the simplified $\mathcal{L}$-transform representation 15.**20-21**.

20 $\mathcal{L}$-*transformation of Model P*

As we have seen in Chapter 13, Model P consisted of the equation

21 $$\dot{q} = (\mathbf{E} - \mathbf{A})r - c$$

of the real processes (13.**12**) and three equations of the control subprocesses (13.**14-16**)

22 $$\dot{p} = -2\beta\gamma\dot{q} - \gamma^2(q - q^*) \qquad \text{(price setting)}$$

EQUIVALENCE OF CONTROLLERS

23 $\quad g = (E - A')p \quad$ (profitability calculation)

24 $\quad \dot{r} = \alpha \dot{g} \quad$ (production decision)

22-24 could have been condensed into one equation (13.19)

25 $\quad \dot{r} = \alpha(E - A')[-2\beta\gamma\dot{q} - \gamma^2(q - q^*)].$

The additional equation (13.10)

$$Y = A\langle r \rangle$$

can be left out of consideration in what follows because if the r paths are identical in the two models, the Y paths will be also.

Let us now take the Laplace-transform of equations **21** and **25** by the use of Tables 5 and 6:

26 $\quad s\,q - q_0 = (E - A)r - c$

27 $\quad s\,r - r_0 = (E - A')[-2\beta\gamma(s\,q - q_0) - \gamma^2(q - \dfrac{1}{s} q^*)].$

Let us now define new variables $\bar{q}(s)$, $\bar{r}(s)$ and $\bar{q}^*(s)$ in the following way:

28
$$\bar{q} := q - \frac{q_0}{s}$$

$$\bar{r} := r - \frac{r_0}{s}$$

$$\bar{q}^* := \frac{\gamma(q^* - q_0)}{2\beta s^2 + \gamma s}.$$

It is to be noted that the coordinate transformation **28** is carried out component-wise, thus no new interdependence among the variables has been produced nor old ones decoupled.

In terms of the new variables (with overbar), equations **26-27** take the following form:

29 $\quad \bar{q} = \dfrac{1}{s} (E - A)\bar{r} + \dfrac{1}{s^2} (E - A)r_0 - \dfrac{c}{s}$

30
$$\bar{r} = -\alpha(E - A')\left(\frac{2\beta\gamma s + \gamma^2}{s}\right)(\bar{q} - \bar{q}^*).$$

(By substituting into **29-30** according to **28** we retrieve **26-27**).

The equations **29-30** already correspond to the simplified $\mathcal{L}$-transform 15.**20-21**, with the following notation (the superscript refers to Model P):

31
$$y := \bar{q}, \quad v := \bar{r}, \quad z := \begin{bmatrix} r_0 \\ c \end{bmatrix}$$

32
$$y^{*P} := \bar{q}^{*P} = \frac{\gamma^P(q^{*P} - q_0)}{2\beta^P s^2 - \gamma^P s}$$

33
$$H^P := \frac{1}{s}(E - A), \quad I^P := [\frac{1}{s^2}(E - A) \vdots -\frac{1}{s}E]$$

34
$$M^P := -\frac{\alpha(2\beta^P\gamma^P s + \gamma^{P2})}{s}(E - A').$$

35 **$\mathcal{L}$-transformation of Model SM**

Following the same steps as in paragraph **20** we get the $\mathcal{L}$-transform of the equations of Model SM:

36
$$\dot{q} = (E - A)r - c$$

37
$$\dot{r} = -2\beta\gamma\dot{q} - \gamma^2(q - q^*)$$

in the form

38
$$s\,q - q_0 = (E - A)r - c$$

39
$$s\,r - r_0 = -2\beta\gamma(s\,q - q_0) - \gamma^2(q - \frac{1}{s}q^*).$$

After the coordinate transformation **28**, equation **38** takes the same form as **29**, while **39** gives:

$$\bar{r} = \frac{2\beta\gamma s + \gamma^2}{s} (\bar{q} - \bar{q}^*).$$

Hence the comparison with the simplified $\mathcal{L}$-transform representation yields the same variables **y**, **v**, **z** as in **31** while for **y***, **H**, **I** and **M** we get

40
$$\mathbf{y}^{*SM} := \bar{\mathbf{q}}^{*SM} = \frac{\gamma^{SM}(q^{*SM} - q_0)}{2\beta^{SM}s^2 + \gamma^{SM}s}$$

41
$$\mathbf{H}^{SM} := \frac{1}{s}(\mathbf{E} - \mathbf{A}), \quad \mathbf{I}^{SM} := [\frac{1}{s^2}(\mathbf{E} - \mathbf{A}) \vdots -\frac{1}{s}\mathbf{E}]$$

42
$$\mathbf{M}^{SM} = -\frac{2\beta^{SM}\gamma^{SM}s + \gamma^{SM2}}{s}\mathbf{E}.$$

43 *Testing the equivalence*

The two systems $\mathcal{S}^P$ and $\mathcal{S}^{SM}$ belong to the same basic set $\mathcal{H}$, since both could be represented as a simplified $\mathcal{L}$-transform and in this form we had

$$\mathbf{H}^P = \mathbf{H}^{SM} \quad \text{and} \quad \mathbf{I}^P = \mathbf{I}^{SM}.$$

Since the **I** matrices have full column rank and the **H** matrices full row rank, Corollary **14b)** applies, and the equivalence conditions **17-18** yield

44
$$\mathbf{M}^P = \mathbf{M}^{SM}$$

45
$$\mathbf{M}^P \mathbf{y}^{*P} = \mathbf{M}^{SM} \mathbf{y}^{*SM}$$

The condition **45** is not worth dealing with because already condition **44** cannot be satisfied. Namely, on the left hand side we have the irreducible matrix $\mathbf{M}^P$ (a scalar multiple of $\mathbf{E} - \mathbf{A}'$) while on the right hand side we have the diagonal matrix $\mathbf{M}^{SM}$ (a scalar multiple of the unit matrix). These cannot be equal except for the irrelevant one-sector case and the case $\gamma^P = \gamma^{SM} = 0$, excluded long ago.

Thus Model P and Model SM cannot be equivalent for any value of the parameters appearing in them. Even if we replace the scalar parameters β and γ by (*diagonal,* nonsingular) matrices this consequence cannot be changed.

In looser economic language the above result can be interpreted in the following way: if we base the production control on *output-stock signals*, or *supplier's price signals* instead, the time paths of the production level and output stock will be necessarily different, even if initial values, stock norms and end uses are the same in both systems.

46 Well, what of it?

The reader who has followed my line of reasoning thus far may be indignant on two accounts:
— Why had the above negative statement on the equivalence of two models to be asserted at all?
— The final conclusion could already have been drawn from collating equations 25 and 37. Why were we forced to struggle through all these irksome mathematical manipulations?

The answer to these questions is also twofold:
— The negative result serves as a starting point for looking for other price-setting principles beyond those applied in Model P, if I wish to attain equivalence. To this end, the formalization which I have introduced will be of great help, and would in any case have proved necessary sooner or later.
— I have also given a foretaste of the idea that in the equivalence analysis the structure (diagonality, irreducibility) of the control matrix **M** plays a crucial role.

This is what I am going to discuss in the next chapter.

Chapter 17

PARTIALLY COORDINATED EQUIVALENT CONTROLLERS (MODEL E)

1 *The order of the discussion*

In the preceding chapter we saw that the production control based on prices set by the supplier and on profitability calculations cannot be equivalent to that based purely on output-stock signal. The question arises now whether, by applying some other principle of price setting and perhaps other production decision pattern, we shall be able to construct a controller which is already equivalent to the pure output-stock controller. (Anticipating the answer: yes, but such a control implies coordination.)

But here I will break with the barely justifiable practice of constructing models for comparative purposes on the spur of the moment. Rather I am going to select from a wide spectrum of controllers with prespecified characteristics all those which are equivalent to Model SM, and classify them according to structural criteria.

2 *Notation*

For this study a simple theorem from matrix algebra will be needed. In this theorem the *square* matrices D, R, B, P are not necessarily time dependent in contrast with our accepted notational rule, nor should they be identified with matrices denoted by the same letters in other chapters.

3 Theorem on the diagonality of treble matrix products

Let D be a diagonal non-singular matrix, and R, B, P be square matrices such that $D = RBP$.

a) If B is non-diagonal, then at least one of the matrices R and P is non-diagonal.

b) If B is irreducible and either R or P is diagonal, then the other is irreducible.

4 Proof

a) Since D is non-singular, R and P must be non-singular as well, hence $B = R^{-1}DP^{-1}$. If R and P were both diagonal then on the right hand side we have a product of three diagonal matrices which could not equal the non-diagonal B. Hence at least one of R and P is non-diagonal.

b) Assume for instance that R is diagonal and for the sake of contradiction that P is reducible. The reducibility of P means that (perhaps after symmetrical permutation of rows and columns) it can be partitioned in the form

$$P =: \begin{bmatrix} P_{11} & P_{12} \\ 0 & P_{22} \end{bmatrix},$$

where P_{11} and P_{22} are non-empty square matrices not necessarily of the same order. Let us partition B and $R^{-1}D$ conformably, taking into account that $R^{-1}D$ is diagonal.

$$B =: \begin{bmatrix} B_{11} & B_{12} \\ B_{21} & B_{22} \end{bmatrix}, \qquad R^{-1}D = \begin{bmatrix} R_1^{-1}D_1 & 0 \\ 0 & R_2^{-1}D_2 \end{bmatrix}$$

Since we must have $BP = R^{-1}D$, we get $B_{21}P_{11} = 0$. But P_{11} cannot be singular since in that case P would be also, thus $B_{21} = 0$ in contradiction with the assumption that B is irreducible. Hence P must be irreducible, if R is diagonal. The case when P is assumed to be diagonal can be treated similarly.

5 Subprocesses in the controller of Model P

We have seen under 16.22-24 that in Model P the control of production divides into three subprocesses: price setting, profitability calculation and production decision. If we carry out the Laplace transformation on these equations one by one, introduce the new variables (with overbar) as defined in 16.28 and analogously:

6
$$\bar{p} := p - \frac{p_0}{s}$$

$$\bar{g} := g - (E - A') \frac{p_0}{s},$$

then we get the three equations in the following form:

7 $$\dot{\bar{p}} = -\bar{P}(\bar{q} - \bar{q}^*) \qquad \text{(price setting)}$$

8 $$\bar{g} = (E - A')\bar{p} \qquad \text{(profitability calculation)}$$

9 $$\dot{\bar{r}} = \bar{R}\bar{g}, \qquad \text{(production decision)}$$

where

10 $$\bar{P} := \frac{2\beta^P \gamma^P s + \gamma^{P2}}{s} E$$

is called the *pricing matrix* and

11 $$\bar{R} := \alpha E$$

the *production decision matrix*. It can be seen that in this way the transfer matrix M^P of the controller (cf. 16.34) has been produced in the form:

12 $$M^P = -\bar{R}(E - A')\bar{P}.$$

13 Constructing Model E

Now we are going to construct Model E (E for equivalent) in the following way.
— We leave unaltered everything which was common to Model SM and Model P, this is to say the real processes, and the fact that the vector q (or after the Laplace transformation the vector **q** as defined in 16.**28**) is what is measured.
— We insist on somehow forming a price vector p and from it the value-added vector g with the help of the matrix $(E - A')$. After the Laplace transformation we get the vectors $\bar{\mathbf{p}}$ and $\bar{\mathbf{g}}$ connected by equation **8**.
— We reject however the specification **10-11** of the matrices $\bar{\mathbf{P}}$ and $\bar{\mathbf{R}}$ and replace them with unspecified matrices **P** and **R** (dependent on s) representing the new pricing matrix and production decision matrix for Model E. These matrices may have any structure in contrast with the diagonal matrices $\bar{\mathbf{P}}$ and $\bar{\mathbf{R}}$.
— Finally: among the controllers defined in this way we look for those which are equivalent to the controller of Model SM.

Thus Model E is specified in the following form:

14 $$\bar{\mathbf{q}} = \frac{1}{s}(E - A)\bar{\mathbf{r}} + \frac{1}{s^2}(E - A)r_0 - \frac{c}{s}$$

15 $$\bar{\mathbf{r}} = -R(E - A')P(\bar{\mathbf{q}} - \bar{\mathbf{q}}^{*E}).$$

In this way we obtain

16 $$H^E = H^P = H^{SM}, \qquad I^E = I^P = I^{SM}$$

so that Model E is comparable with Model SM and

17 $$M^E := -R(E - A')P$$

18 $$y^{*E} := \bar{\mathbf{q}}^{*E}.$$

In this representation **R**, **P** and $\mathbf{q}^{*E}$ are as yet unspecified; they will be characterized by the equivalence conditions.

19 *The analysis of equivalence*

Let us compare now the two models SM and E both having been put in the simplified $\mathcal{L}$-transform representation, the Model E specified in **16-18** and Model SM in 16.**40-42**. Since the common **H** and **I** matrices have full row and column rank respectively, we can apply the equivalence criteria 16.**17-18** to get:

20 $$R(E - A')P = \varphi(s)E$$

21 $$R(E - A')P q^{*E} = \varphi(s)\bar{q}^{*SM},$$

where

22 $$\varphi(s) := \frac{2\beta\gamma s + \gamma^2}{s}$$

is a scalar function of s and the superscripts SM were suppressed from β^{SM} and γ^{SM}.

If **20** holds then from **21** in view of 16.**40** we get

23 $$q^{*E} = \bar{q}^{*SM} = \frac{\gamma(q^{*SM} - q_0)}{2\beta s^2 + \gamma s}$$

If we substitute into the equation

$$\bar{r} = M^E(\bar{q} - q^{*E})$$

according to 16.**28**, 17.**17**, **20** and **23** and perform the inverse Laplace transformation, we conclude that

24 $$q^{*E} = q^{*SM}.$$

This is to say that the models E and SM cannot operate equivalently unless their output stock norms are identical. This is not a surprising result if we consider the fact that in the stable case (stability was not assumed!) q^{*SM} is the steady state of q in Model SM, hence for identical q paths it cannot be otherwise in Model E either.

Indeed, it is not this but equation **20** that deserves more of our attention.

25 Three structural variants of Model E

Let us observe first of all that the matrix $\varphi(s)E$ on the right-hand side of **20** is diagonal and nonsingular, and that the middle factor $(E - A')$ on the left-hand side treble product is irreducible (*a fortiori* non-diagonal) by assumption. Thus Theorem **3** applies: at least one of the matrices **R** and **P** is non-diagonal, and if either is diagonal, the other is irreducible.

From this we conclude that there are three collectively exhausting and mutually exclusive structural variants of Model E:

Variant EP: **R** is diagonal, **P** is irreducible

26   Variant ER: **P** is diagonal, **R** is irreducible

Variant EM: both **R** and **P** are non-diagonal. (M for mixed)

This being said, we can start to interpret these three variants in economic terms.

27 Stability and viability

The controller of Model E is equivalent to that of Model SM, thus if the latter is stable and viable so is the former. Although the stability (and the concluding viability) conditions have not been formally stated for Model SM, this was done for Model S from which it was derived by a simplifying assumption. Thus there is no reason to detail here this stability analysis, already routine, and which would only yield similar results to those obtained in Chapter 9 (with a different lower bound on β) without providing any new lessons.

28 Preliminary remarks

Let us recall first that the three variants which were specified in **26** by their dissimilarities agree with each other at every other point. In particular, their real processes are the same, as are the measurement processes, in that the actual output-stock q is observed and compared with the same normal stock q^*. (See **24**.) They also agree in that prices are set and from them profitability indicators $g = (E - A')p$

calculated, which figure in the control of production. Decision on transfers is also made in the same way. (To this I will return in a moment.) The difference is thus restricted to the question: how are the prices (**P**) set and how are the decisions on production (**R**) made? What is the structure of the information and decision processes connecting them? How can these structures be interpreted in economic terms?

The natural, logical order of succession of the subprocesses is the following:

measurement of the output stocks $\rightarrow$ price setting $\rightarrow$

profitability calculation $\rightarrow$ decision on production $\rightarrow$

decision on material purchases.

It is, however, reasonable to carry out the structural analysis in reverse order, to start with the purchases and end up with the measurement of the stocks. This backtracking can be defended by the following reasoning. We cannot tell whether a subprocess is communicative or not unless we have already ascertained whether the next one is based on information coming from outside. Thus, if we wanted to keep the logical sequence of subprocesses, we could not finish the analysis of one subprocess without having analysed the subsequent one, i.e. without jumping ahead and back. This is why it is expedient to reverse the direction of the study, and I will restore the logical order when summarizing each variant.

29 *The purchasing decision*

This subprocess is the same in all the three variants, inherited from Models SM and P. I am speaking here about the equation

$$Y = A<r>$$

or, more strictly, about its $\mathcal{L}$-transform which has not been explicitly discussed. Let me do this here entry by entry:

$$\mathbf{Y}^{ij} = \mathbf{A}^{ij}\mathbf{r}^j. \qquad \forall i,j$$

In Model E (as well as in models SM and P) transfers are decided by the buyer j who is supposed to know both his own technological coefficients A^{ij} and the production decision r^j (whether decided by himself or transmitted to him). Thus the producer of a commodity can decide on his purchases *without coordination* and can carry out the purchase by himself *without communicating* his decision to anybody else. (It is not to be communicated even to the seller, who does not observe sales directly, but only the resulting level and change of stocks.)

Since the above statement holds true with respect to all three variants under study, I will not repeat it in the course of their piecemeal analysis.

30 Variant EP: coordinated price setting

The matrix **R** of production decision, which precedes the purchasing decision, is diagonal in variant EP, thus the volume of production of each commodity depends only on the corresponding component of the indicator g. In the knowledge of this single value the producer can decide on the production level in an uncoordinated way, and need not communicate his decision to any outside organization, thus *the production decision subprocess is non-communicative and uncoordinated*.

In order to calculate a component of the vector g the producer must know the price of his own product and of all the materials he uses as inputs, and presumedly he knows his own input coefficients, a column of **A**. The profitability calculation can thus be carried out without coordination and its result need not be communicated to anybody else. Thus the *profitability calculation subprocess* is also *non-communicative* and *uncoordinated*.

The pricing matrix **P** is irreducible. This means that in order to set any one price, one needs information (concerning the deviation between actual and normal stocks) stemming from a wide range of producers. No price can be formed by the producer in an uncoordinated way, price setting must be coordinated. And the prices set in any coordinated way must also be communicated not only to the producer of the commodity but also to all those who use the commodity as input. Thus *the price setting subprocess* is *communicative and coordinated*.

We have seen under 4.4 that a coordinated subprocess may be one of two kinds: centralized or interactive. Accordingly the price setting subprocess may be given two alternative interpretations.

31 First interpretation: centralized price setting in variant EP

One of the possible interpretations of coordinated price setting is that it is done by a control organization set up for this purpose and appropriately called *Price Office*. In possession of the necessary information it sets the prices in a centralized way and communicates them to the interested producers. Thus in this interpretation the *price setting subprocess* is *communicative and centralized*, and the communication is *non-transactional* since a control organization is the sender.

The centralized price setting process is preceded by the measurement process which is *introspective* according to Assumption 4.9, the output stock of a commodity is observed only by its producer. Then the output stock data which is the input signal of the price setting process are to be transmitted to the Price Office. Thus the *measurement process*, besides being *introspective*, is also communicative, and since the addressee is a control organization, *non-transactionally communicative*.

32 Second interpretation: interactive price setting in variant EP

Another possibility for interpreting coordinated price setting is to assume that no separate control organization (Price Office) takes part, but that the prices are set jointly by the buyers and sellers appearing on the market in an *interactive* way. We cannot describe the inner structure of the interactive price setting process: it must be considered as a *black box*. Its input signal is the difference vector between the actual and normal output stocks, the output signal is the price vector, and within the box information and decision processes are enacted of which we are ignorant, save that we know there is no other participant in the process than the real organizations, the producers. The prices, which appear at the end of the price setting process, are known by all participants, thus the process is communicative. The structure of matrix $\mathbf{P}$ is identical with that of $\mathbf{C}' = (\mathbf{E} - \mathbf{A}')^{-1} > 0$ as can easily be seen. Hence $\mathbf{P}$ contains no zero entry and the indirect material input relations become direct in the course of the price setting process, pairs of producers become connected even if there is no direct transaction between them. Thus the *interactive price setting* is *non-transactionally communicative*.

It is a long established tradition in economics to conceptualize the price setting function of the market in terms of the notions "supply" and "demand" or "oversupply" and "overdemand". In my models there are no variables which could be directly identified with these concepts. However, in order to bring the previous description of price setting closer to the customary realization of the market, I suggest the following interpretation. The output stock is nothing other than the cumulated difference between past production and sales. If we identify the amounts produced as supply flow and the amounts sold as demand flow (this being justified in the case of invariably non-negative stocks), then actual output stock can be interpreted as "cumulated oversupply" and its derivative as "oversupply flow". Thus the market in my description can be interpreted as changing prices in response to that part of cumulated oversupply above (or below) the normal level and to the instantaneous oversupply flow in such a way that in the formation of any one price the oversupply data of all commodities must be taken into account.

After this digression let us return to the point where we left off the interpretation of variant EP in the case of interactive price setting. The measurement process is, of course, *introspective* anyway, but since we have no inspection into the price setting black box, it does no harm to assume that the information exchange happens in the course of price setting and not before it. In this way the measurement process becomes *non-communicative*.

33 Variant EP: summary of the interpretation

In variant EP the price setting process is coordinated and communicative, the coordination may either be centralized (in the Price Office) or interactive (on the market). The preceding measurement process is non-transactionally communicative in the first case, and non-communicative in the other. The profitability calculation is uncoordinated and non-communicative as is the final decision on production and purchases.

The distinctive feature of variant EP is the coordinated price setting and the uncoordinated production decision.

34 Variant ER: coordinated production decision

In this case the matrix **R** of production decisions is irreducible; thus the control of production is necessarily coordinated in this variant. In view of our scheme two interpretations are possible depending on whether we consider the subprocess to be interactive or centralized.

The interactive case, however, could only be given a quite arbitrary interpretation. Namely, we had to assume that the producers knew the profitability of each other's products, all the components of the indicator g, and collectively decided obligatory production quotas for each producer. Such an organization (pool) is known to occur within some industries (OPEC is the most widely known example) but not across industries. If some interactive coordination of production plans still occurs, e.g. in the form of long-term contracts, then its purpose is to assure some supply-demand equilibrium in real quantities rather than an interactive decision based on profitability considerations. By this reasoning we refrain from interpreting variant ER as an interactive production decision process.

Thus in this variant the production decision is *centralized* in the hands (say) of a *Production Office*, which sets the level of production for each producer on the basis of the profitability calculations. Since this decision must be transmitted to the producers who effect it, the production decision is *non-transactionally communicative*.

The preceding process, the profitability calculation, may however, be interpreted in two alternative ways; the technology may be public (i.e. known to the Production Office) or secret (known only by the individual producers).

35 First interpretation: public technology in Variant ER

Here we assume that the Production Office knows the whole input coefficient matrix **A**. (Since **A** is constant in time, it may have become known by a process outside the current dealings.) In this case the whole vector g can be calculated in the Production Office in the knowledge of the price vector p. Thus *the profitability calculation* is *centralized* and, since the signal g is used only in making decisions within the Office, it is also *non-communicative*.

The pricing matrix **P** is diagonal, thus the *price setting* process is *uncoordinated* and is *non-transactionally communicative* since the prices are to be transmitted to the Production Office.

36 *Second interpretation: secret technology in Variant ER*

In this case each producer knows (and only he knows) his own technology, which is a column of **A**. From this he can calculate one component of g, the profitability of his product, if he knows the current price vector p. Thus in this case the *profitability calculation* is *uncoordinated*, but the profitability index is to be transmitted to the Production Office, hence the subprocess is *non-transactionally communicative*.

The *price setting* process is, of course, uncoordinated in this interpretation also but now it is *transactionally communicative* since each price is to be made known to all users of the commodity.

In both interpretations *the measurement process* is not only introspective but also *non-communicative* since each price depends on the corresponding output stock only.

37 *Variant ER: summary of the interpretation*

In variant ER the decision and information structure is the following.
— The measurement process is introspective and non-communicative, the price setting process is uncoordinated and communicative. This communication is non-transactional if the technology is public and transactional if it is secret.
— The profitability calculation is uncoordinated and non-transactionally communicative (in the first case), and is centralized and non-communicative in the second case.
— The production decision is centralized and non-transactionally communicative.

The distinctive feature of variant ER is the centralized production decision and uncoordinated price setting.

38 The duality between EP and ER

I will make a short digression here: the conspicuous duality between the two variants seems to be worth discussing. With the former we have met coordinated price setting and uncoordinated production decision, with the latter this was reversed. We find an evident analogy with long established duality concepts in other types of decision processes between the handling of quantity-type and price-type information. I refer, for instance, to different decomposition principles in mathematical programming.

The primal decomposition principle implies that the center (the master programme) transmits price-type (dual) signals and the sectors send quantity-type signals to the centre. This recalls variant EP. The dual decomposition principle reverses the direction of the signals, as does ER. It was once suggested that the first principle corresponds to an abstract picture of the centrally planned economy and the second to that of a market economy. Our analysis has shown that the parallelism is valid only with respect to the information structure, and cannot be extended to the (price) signal generation, interpretable both as the result of a market process and as a centralized process.

39 Variant EM: the mixed case

In this variant the only thing we know is that neither the pricing matrix **P**, nor the production decision matrix **R** is diagonal. From this no other conclusion can be drawn than that *there are* certain commodities whose price or production level is set in a coordinated way, and at the same time there *might be* prices and/or production levels set by the producers without coordination. The organizational structure may contain disconnected partial markets, separate Price Offices and Production Offices, whose authority is restricted to a subset of the commodities. This is thus a mixed system in which the hitherto pure types of price setting and production decision processes appear side by side. Within groups of commodities prices are coordinated but across these groups they are not (and a single commodity may itself form such a group) and the same applies to the production decision process. Moreover, the classification of commodities in one subprocess is — as a rule — different from the classification in the other subprocess.

We conclude that the distinctive characteristic of Variant EM is that the price setting process and the production decision are both partially coordinated.

It is quite obvious that variant EM is the one closest to the information and decision structure of actually existing economies. No wonder that on the present level of abstraction and generality no further lesson could be drawn from its study. Let us thus turn our attention to a special EM structure.

40 A particular EM structure

Let us divide the commodities into two groups. In the first group (intermediate and final products, say) prices are set by the manufacturers in an uncoordinated way. In the second group (raw materials) price setting is coordinated by the Price Office (or, for that matter, by the Commercial Exchange).

The following questions can now be asked:
— Is it possible to achieve an arrangement in which the production of the raw materials is decided by the producers in an uncoordinated way and only the production decision concerning intermediate and final products needs coordination?
— If the answer is yes, does this arrangement imply a saving in information transmission and processing in comparison with other arrangements? (Or to put it more bluntly: Can partial coordination of a subprocess be efficient?)

The answer is affirmative to the first question and negative to the second.

Let us formulate the equivalence condition **20** in a partitioned form corresponding to the above classification (the notation is self-explanatory).

41
$$\begin{bmatrix} R_{11} & R_{12} \\ 0 & \hat{R}_{22} \end{bmatrix} \begin{bmatrix} E_1 - A'_{11} & -A'_{12} \\ -A'_{21} & E_2 - A'_{22} \end{bmatrix} \begin{bmatrix} \hat{P}_{11} & 0 \\ P_{21} & P_{22} \end{bmatrix} = \varphi(s) \begin{bmatrix} E_1 & 0 \\ 0 & E_2 \end{bmatrix}.$$

Here the nonsingular diagonal matrices $\hat{P}_{11}$ and $\hat{R}_{22}$ can be chosen arbitrarily, and thus we get a unique solution for R_{11}, R_{12}, P_{21} and

P_{22}. The scalar factor $\varphi(\mathbf{s})$ is irrelevant for the present analysis and is thus omitted.

$$\mathbf{R}_{11} = \hat{\mathbf{P}}_{11}^{-1}[\mathbf{E}_1 - \mathbf{A}'_{11} - \mathbf{A}'_{12}(\mathbf{E}_2 - \mathbf{A}'_{22})^{-1}\mathbf{A}'_{21}]^{-1}$$

$$\mathbf{P}_{22} = (\mathbf{E}_2 - \mathbf{A}'_{22})^{-1}\hat{\mathbf{R}}_{22}$$

$$\mathbf{R}_{12} = \mathbf{R}_{11}\mathbf{A}'_{12}(\mathbf{E}_2 - \mathbf{A}'_{22})^{-1}$$

$$\mathbf{P}_{21} = (\mathbf{E}_2 - \mathbf{A}'_{22})^{-1}\mathbf{A}'_{21}\hat{\mathbf{P}}_{11}.$$

42

From the properties of matrix **A** it is not hard to deduce that the matrix inverses on the right hand sides exist, thus **41** has a solution. This affirms a positive answer to the first question: the arrangement is realizable.

Concerning the second question it is to be observed that neither $\mathbf{R}_{12}$ nor $\mathbf{P}_{21}$ can be a zero matrix; thus neither the pricing matrix nor the production decision matrix can be put in a block diagonal form. A glance at **41** reveals the lesson: for the centralized price setting of raw materials the Price Office requires information stemming from the uncoordinated sector, and for the centralized decision on the production of intermediate and final products the *Production Office* must collect information from the raw material sector. Thus there is hardly any saving in the amount of information to be transmitted compared with a fully centralized arrangement both on the pricing and on the production side; there is, however, a marked loss in comparison with the two pure arrangements ER and EP. Hence the negative answer concerning the informational efficiency of the partial coordination in each subprocess.

Although the above arguments rely on particular equivalence conditions, I will risk drawing a more general lesson. We have observed in real life how a control organization, whose authority is restricted to a segment of the economy, tends to extend its authority to outside areas. In the light of our analysis this tendency is reasonable since the fulfillment of its duty requires a vast amount of data from without. *It may be more efficient to centralize the control of a few subprocesses of the economy forcefully, leaving the rest alone, than to scatter limited central controllers across the subprocesses.*

Chapter 18

SUMMARY OF PART THREE

1 *Contrasting Part Three with Part Two*

In Part Two, I showed that the same real sphere can be operated by different *vegetative* controllers which do not require any coordination, and where either no communication among the agents is needed or else there is some transactional communication using quantity (order) or price signals. These different controllers produced different trajectories for production, stocks and transfer of commodities.

In Part Three, on the other hand, I generated controllers such that the operation of the systems were equivalent in the sense of identical trajectories. These systems, however, were not vegetative any more, but required *coordination* in one part or another of the multistage control process, and implied different types of information flow patterns. For their operation specialized control organizations were needed as being separated from the real organizations (productive firms) or alternatively the market had to be operating as providing the necessary information link among them and playing a price setting role.

In this sequel I will try to summarize in a few sentences the economic lessons to be drawn from the study carried out in Part Three. (Although I will focus my attention on these results I will glance back at Part Two here and there.) In the course of doing so I will omit the listing of those conditions and assumptions which I think do not pertain to the essence of the economic process under study but serve only to facilitate the mathematical analysis. Thus my proposition is more general and less exact in comparison with what could be deduced from the models by rigorous formal reasoning.

2 Characterization of the economy under study

My propositions refer to an economy with the following characteristics:

a) Real activities are carried out only by productive firms and end users in the economy. (In one model of Part Two a trader has been added.)

b) Production is restricted by the existing productive capacity. Beyond this restriction, both the scarcity of primary resources and their allocation is neglected.

c) Financial (budget) constraints do not restrict the purchases of either the producers or the end users.

d) Changes in technology and consumption patterns are relatively slow compared to the speed with which information is acquired, transmitted and processed, and decisions carried out.

e) The producers (and, if they appear, traders and control organizations) form their decisions on production level, purchases and deliveries so as to comply with certain simple behavioural rules. (The same is not assumed of the end users.) This applies also to the generation of such interior information as prices, profitability indicators or order quantities.

These assumptions, of course, severely restrict the scope wherein my models can be considered as an approximate description of real-world economies. Individual models represent a particular aspect of the economy at most. Collectively they do this for a set of phenomena which appeared worth studying for their importance.

Subsequent analyses should and hopefully will progress in the direction of extending to the resource allocation mechanism, technical progress, monetary processes and institutions, as well as to a better representation of the motivation of the economic agents.

3 The viability of the economy

From the study of the economy as characterised above I draw the following conclusions: The economy can be operated by the use of *different economic mechanisms* so that it would resist a restricted amount of external shocks (disturbances, loads). The most important point was to show how limited an amount of coordination and information flow is required for the survival of the economy.

4 The variety of economic mechanisms

In Part Two the variety of economic mechanisms was due mainly to the diversity of the *contents* of the control signals (stocks, orders, prices), from which different information structures followed for controllers not containing any coordinating organisation.

In Part Three the following sequence of signal generation was studied:

output stock → price → profitability indicator →

production decision → purchasing decision.

From this I deduced different *organizational* structures: market(s), Price Office(s), Production Office(s) and corresponding patterns of information flow. These organizational variants, as well as their interpretations are summarized in Table 7.

5 Alternative interpretation possibilities

The three model variants EP, ER and EM (all equivalent to each other and to SM) were proven to exhaust all possibilities. However, some of them admitted alternative interpretations, and I have never suggested that these shown in Table 7 likewise exhaust all the alternative interpretation possibilities. On the contrary, I strongly suggest that there might be others. For instance, there is nothing to prevent an organization from starting to control any decision process (or more generally: signal transformation) which the real organizations were originally able to control in an uncoordinated way. It seems that coordinatedness has a lower bound only (which I tried to explore in Chapter 17), but no upper bound. Real life observations of the expansive tendencies in economic bureaucracy seem to support this lesson.

Table 7

Mark of the model variant	SM	EP	ER	EM
Main characteristic	Uncoord.	Coord. price setting	Centralized production decision	Part. coord. price or production decision: policentric
Alternative interpretations	— —	Centralized: Price Office	Public technology	— —
		Interactive: "market"	Secret technology	
Measurement of stocks; introspective and	Non-com.	Com.	Non-com.	Part. Com.
		Non-com.		
Price setting	None	Centralized, Com.	Uncoord. Com.	Part. Coord. and Com.
		Interactive Com.	Uncoord., Trans. Com.	
Profitability calculation	None	Uncoord., Non-com.	Centralized Non-com.	Uncoord. or Part. Centralized
Production decision	Uncoord., Non-com.	Uncoord., Non-com.	Centralized, Com.	Part. Centralized and Com.
Purchasing decision	Uncoord., Non-com.	Uncoord., Non-com.	Uncoord., Non-com.	Uncoord., Non-com.
Control organization	None	Price Office	Production Office	Market(s), Production Office(s), Price Offices(s)
		Market		

Abbreviations
Com. := Communicative
Coord. := Coordinated
Part. := Partially
Trans. := Transactionally

6 Indifference to history?

In the course of the analysis in both Parts Two and Three it somehow remained unclear in which historical era certain aspects of economic behaviour and control structures can be linked to those represented by my models. On the one hand it is clear that the real sphere of these economies is based on a well-developed division of labour, consisting of specialized economic agents who perform different productive activities. From this aspect one is inclined to think of the models as representing "modern" economies as we encounter them in contemporary capitalistic and socialistic societies.

As compared with this approach to the real sphere, the representation of the control mechanisms seems to be rather primitive, more reminiscent of precapitalistic or even more ancient economic formations. The use of the terms "market", "buyers and sellers", were also misleading to some extent, serving only to familiarize the reader with my reasoning without burdening him with undigestible terminology.

It is not only the omission of monetary processes which poses this question, but rather the omission of "exchange of commodity for commodity in an agreed proportion". What is actually represented is—in Marxist terms—rather an "exchange of activities" than anything else. From a historical point of view this arrangement most resembles a stone age tribal economy, where everything produced by tribesmen of differing skills is shared by the community so as to guarantee the survival of the tribe. The observation of the depletion and level of stocks (control by stock norms), the communication of orders fulfilling future needs (e.g. of stone axes), even some coordination of the tribe's activities by the chieftain, the medicine man or the elders of the people, can be well interpreted in this stone age setting.

How can the contradiction between modern industrial technology and stone age control of the economy be reconciled? How can the indifference to history be explained and justified? The answer is twofold.

7 The stone age is present

What I called above stone age control of the economy is not extinct. It is true that contemporary economic control is much more refined, and has a more complex structure. A series of economic behavioural rules and organizations was born and died in the course of the economic history of our society, and the way the economy operates today cannot be explained *merely* by the set of models I have studied. But the rules followed by our ancestors are still with us. No specific economic system can be well understood if we are not able to separate that part of the structure which distinguishes it from other — past, present and potential future — economic systems, from the part which they have in common. It is, indeed, surprising how large a part of the working of present day economy can be described in terms which are interpretable also in stone age relations, without much arbitrariness.

Let me draw an analogy from biology. The life form of a mammal is very different from that of an earthworm. Even so, many basic functions of both (such as metabolism, karyologic processes etc.) are controlled in a similar way. No biologist would doubt that the higher order functioning of the former could not be understood if the structures they share had not been clarified. And I would suggest that contemporary economic structures are less different from stone age structures than a mammal is different from a earthworm, if you do not take the comparison of incommensurabilities seriously.

8 The sequence of models vs historical chronology

The sequence of the models in Parts Two and Three followed a logical order from simpler models to more complicated ones, from less coordination and information flow to a gradually increasing amount of both. But this sequence also reflects a very perfunctory historical chronology — with wide hiatuses and inversions. Within Part Two the progress from completely isolated agents to increasingly communicative ones, the emergence of trade as the activity of specialized agents and the introduction of prices as a first step to a monetarized economy, all reflect stages in the evolution of economic control systems in chronological order. The same is true of the leap from Part Two to Part Three, with the extension of the role of the market and the introduction of centralized coordinating organizations, with such modern concepts as a Price Office.

I am well aware that even the models in whose interpretation the most modern concepts were used have one foot still in the stone age, primarily in the omission of the monetary sphere. However, I hope that their conceptual foundation is strong enough so that the study of economic control structures can be extended to the phenomena of contemporary economy without that unfortunate omission. Only after filling such gaps will it be possible to assess how far indifference to history is justified, and how models should be diversified so as to represent past, present and future real-world economic control systems.

Chapter 19

AFTERWORD:
EQUILIBRIA WITH RATIONING RECONSIDERED

1 *Preliminaries*

In Chapter 2 I drew a rather rudimentary comparison between some of the *starting points* of my approach and the usual assumptions made in the world of General Equilibrium Theory and in particular the Theory of Equilibria with Rationing. I promised there to elaborate on this question later when the value of the different premises could be assessed in the light of the outcomes of the analysis. Now the time has come to fulfil this promise.

However, I encounter my first serious difficulty when trying to define exactly what I am comparing with what. On one side there are enough books and papers on General Equilibrium Theory (GET) to fill a small library, and a smaller but still considerable number of contributions to the Theory of Equilibria with Rationing (TER). On the other side we find the present slender treatise, its predecessors like Kornai's Anti-Equilibrium (AE), the set of papers collected in Non-Price Control (NPC), and a series of more recent papers by Kornai, Simonovits and others more or less related to a similar line of reasoning. If I am not to write another book (still incomplete) on what has been achieved on both sides, I must both be very selective and assume some familiarity on the part of the reader with at least GET.

With such selectivity I can hardly counter the charge of partiality not only in my argumentation (which can be underrated by the reader) but also in neglecting such results as would weaken my position (and such omissions might go unnoticed by the reader). I will try to avoid these biases by selecting contributions of respected advocates of GET and TER and relying more on their self-criticism than on my interpretation.

With respect to GET I rely on the traditional but crystal-clear presentation in the work of ARROW - HAHN (1971): *General Competitive Analysis* and, for recent advances, on FISHER (1983): *Disequilibrium Foundations of Equilibrium Economics*. With respect to TER the main source is BENASSY (1982): *The Economics of Market Disequilibrium* supplemented by MALINVAUD (1977) and the excellent review article by GRANDMONT (1977).

2 General Equilibrium Theory: a bird's-eye view

Generations of outstanding economists have taken pains to make the two century old "invisible hand" of *Adam Smith* more visible and less mysterious. The first major step was taken by *Karl Marx*, whose schema of "simple reproduction" can be regarded as the first model of GET. The concept of general equilibrium became the fundamental concept of a theory in the hands of LEON WALRAS (1874-77). It took about 80 years of effort by economists such as *Clark, Wicksteed, Edgeworth, Pareto* and *Hicks* and such mathematicians as *A. Wald, J. von Neumann* and *S. Kakutani* before a rigorous mathematical foundation of Walras' theory developed into what we call today the *Arrow-Debreu Theory* (ARROW - DEBREU, 1954; DEBREU, 1959)

A Walrasian economic system consists of firms and households. The households acquire revenue from the sale of their endowments (e.g. the labour service they can offer) and within the limits of the revenue they buy consumer goods. The demand of households depends thus on the prices of resources and goods. The households' demand for produced goods and the production functions of the firms determine the supply of the firms as well as their demand for the inputs. The equilibrium prices are defined as those which equalize supply with demand in the market for each particular commodity simultaneously. The crucial question is: under what conditions can the existence of such prices be guaranteed?

No less interesting is the question how equilibrium prices are formed. In the Walrasian terminology the invisible hand, sufficiently omniscient and omnipotent to perform this task, is called the *auctioneer*, and equilibrium prices are attained by the *tâtonnement* (i.e. trial and error) process. The auctioneer is supposed to learn all the agents' intentions to transact if the interim (non-equilibrium) price quotations are in effect. This is the information input by which he might be able to change the price vector so as to attain equilibrium prices. After this tâtonnement process is finished (and only then) can production, transaction and consumption begin.

I must omit here all details concerning the far flung development of GET, such as the alternative sets of conditions under which existence, uniqueness, stability and/or Pareto-efficiency of equilibria could be proved, as well as the different formulae for the tâtonnement process. I am satisfied with pointing out three characteristics of the Walrasian system which were somehow inherited by the Theory of Equilibria with Rationing:

a) The system is essentially static; equilibrium price, equilibrium state are static concepts. Although tâtonnement is described as a dynamic process, we must not forget that real processes stand still in the meantime.

b) Supply and demand depend exclusively on the prices. This one-sidedness will be relaxed in TER.

c) GET claims to deal with a *decentralized* mechanism, where decisions on real activities are taken by the agents. This holds true with respect to some subprocesses of the whole control process, the decisions taken on production and transfer. But this is preceded by a subprocess of *price setting* fully *centralized* in the hands of the auctioneer. (Cf. Variant EP, 17.**31**.) Thus in my taxonomy GET studies a partially centralized control mechanism. In TER coordination will shift over to quantities, showing in this respect a parallel with variant ER (17.**34**) with a difference, however, between coordination of production and that of trade.

3 GET without tâtonnement

Hardly any scholar cultivating GET would have been satisfied with the tâtonnement story. The auctioneer has always been regarded as a rather unfortunate artifact. In successive steps, first trade outside equilibrium was permitted (NEGISHI, 1962; UZAWA, 1962; HAHN, 1962; ARROW - HAHN, 1971), and subsequently extended to include production and consumption (F.M. FISHER, 1976) but still retaining the price adjustment by the auctioneer. Finally, at the cost of introducing some restrictions on the price adjustment proces, *F.M. Fisher* was able to eliminate the auctioneer altogether and replace it with arbitrating agents. His results are summarized in FISHER (1983).

Two important aspects of the above circle of ideas are worth mentioning. The first, to which I too subscribe, is the special emphasis on the stability of the equilibrium, while, as we will see, stability considerations are usually neglected in the TER. The second, and from my point of view less appealing, aspect is its centering around Walrasian equilibrium. This is why I have opted for such a short assessment of the non-tâtonnement literature and will devote more space to TER, even though expecting that the former will yield significant results in non-Walrasian analysis in the future.

4 *The birth of TER*

By the fifties of our century, the period when Walrasian theory had its very foundation, *Keynes* became the mastermind of economists and economic policy makers of the Western world. It was soon recognized that Keynesian macroeconomics could not be substantiated on the basis of Walrasian microeconomics. This perception (CLOWER, 1965; LEIJONHUFVUD, 1968) led to the emergence of many new theoretical approaches, TER being one of them. (Nevertheless the first non-Walrasian equilibrium concept of "general overproduction" appears already in *Marx'* analysis of crisis.) The very success of TER has, indeed, been achieved in the laying of foundations to Keynesian macroeconomics, particularly by the explanation of Keynesian unemployment (BARRO - GROSSMAN, 1971, 1976; MALINVAUD, 1977). Since my reason for paying special attention to TER lies in its relation to my approach, I will not pursue its implications for macroeconomics. (For an excellent review see DRAZEN, 1980.)

As with any other theory, the most conspicuous cognizance (the "visiting card") of TER is a set of concepts particular to that theory. We will acquaint ourselves with three of them:

a) Perceived constraint

b) Effective demand

c) Rationing scheme

Let us take a closer look at each one in turn.

5 *The perceived constraint* (CLOWER, 1965)

If actual prices deviate from equilibrium prices and hence the market is not in Walrasian equilibrium certain intended transactions (sales and purchases) cannot be effected, and run into constraints. As such constraints are perceived by an agent, he will change his demand or supply not only with respect to the commodities concerned but also with respect to the rest of the partial markets. This is the well-known *spillover effect*: rationing on one market spills over to others.

Let us consider a system with n commodities indexed by $h = 1,2,...,n$ and m agents indexed by $i = 1,2,...,m$. The price vector p is assumed to be fixed. The net transaction of agent i is the n-vector y_i, a component y_{ih} of y_i is positive if i is a net buyer of h, negative if he is a net seller.

When the agent visits the markets, he perceives constraints on his transactions, these constraints can be comprised in an n-vector $s_i \leq 0$ of the lower bounds which he meets as seller and an n-vector $b_i \geq 0$ of the upper bounds which he meets as buyer. A component of s_i and b_i is $-\infty$ and $+\infty$ respectively, if i is not constrained in a market as seller or buyer. Agent i possesses a stock-vector $\bar{v}_i \geq 0$ of commodities and an amount $\bar{\mu}_i \geq 0$ of money before trade, and a stock and money holding v_i and μ_i, respectively, after trade, so that

6
$$v_i = \bar{v}_i + y_i \geq 0$$

7
$$\mu_i = \bar{\mu}_i - p'y_i \geq 0.$$

6 is a physical constraint on the non-negativity of stocks, 7 is a financial one, the *budget constraint* of i. The latter stipulates that any actual trade must be backed up by money, the only medium of exchange. The perceived constraints b_i and s_i restrict the transactions:

8
$$s_i \leq y_i \leq b_i.$$

The perceived constraints will be dependent on the rationing scheme in a way to be defined later.

9 The effective demand BENASSY (1977, 1982)

Each agent i is represented by a utility function depending on his holding of commodities and money: $\bar{\omega}_i(v_i, \mu_i)$. Considering **6** and **7**, $\bar{\omega}_i$ can be transformed as depending on y_i only, defining his utility function as

10
$$\omega_i(y_i) := \bar{\omega}_i(\bar{v}_i + y_i, \bar{\mu}_i - p'y_i)$$

since $\bar{v}_i$, $\bar{\mu}_i$, his initial holdings and the price vector p are considered as given data.

The nominal (net) demand of i is defined as the solution y_i^{nom} of the maximization problem

11
$$\max \omega_i(y_i)$$
subject to

12
$$\bar{v}_i + y_i \geq 0$$
$$\bar{\mu}_i - p'y_i \geq 0,$$

but disregarding the constraint **8**. With appropriate smoothness and convexity assumptions on ω_i the problem **11-12** has a unique solution y_i^{nom}.

Let us now define the *effective demand* by taking the perceived constraints **8** into account. If we simply supplement **12** with the constraint **8** (as did DRÈZE, 1975) the agent will be prevented from expressing his wish to sell or buy more than that permitted by constraints **8**. By omitting **8** altogether it would be impossible to take into account the spillover effects. This led *Benassy* to define effective demand of each agent via an n-tuple of maximization problems, each yielding a single component $\tilde{y}_{ih}$ of the effective demand vector $\tilde{y}_i$. In the h'-th problem the perceived constraints **8** $s_{ih} \leq y_{ih} \leq b_{ih}$, $h \neq h'$ are enforced, but $s_{ih'} \leq y_{ih'} \leq b_{ih'}$ is relaxed. Thus for each h' the utility maximization problem takes the following form:

13
$$\max \omega_i(y_i)$$
subject to

14
$$\bar{v}_i + y_i \geq 0$$
$$\bar{\mu}_i - p'y_i \geq 0$$
$$s_{ih} \leq y_{ih} \leq b_{ih} \qquad h \neq h'.$$

From the solution of this problem we pick out $\tilde{y}_{ih'}$ to be the h'-th component of $\tilde{y}_i$ and let h' run over 1,2,...,n. The effective demand vector $\tilde{y}_i$ constructed this way will in general violate not only the constraint 8 but also the budget constraint 7. (**6** remains obviously satisfied.) There is no hitch in this, actual transactions must comply with the constraints, demands need not.

15 *The rationing scheme*

The rationing scheme determines how actual transactions y_i (i = 1,2,...,n) depend on effective demands $\tilde{y}_i$. It is, of course, to be assumed that y_i depends not only on the i-th agent's own effective demand but also on that of all the other agents.

16
$$y_{ih} = f_{ih}(\tilde{y}_{1h'},...,\tilde{y}_{nh}).$$

The rationing scheme is thus represented by the functions f_{ih}.

17 *Properties of the rationing scheme*

Some properties should (or might) be stipulated from a sensible rationing scheme:

(i) On each market actual sales and purchases must balance for whatever effective demands.

18
$$\sum_i f_{ih} = 0, \qquad \forall h, \quad \forall \tilde{y}_{1h'},...,\tilde{y}_{nh}$$

(ii) *Voluntary exchange* (Cf. 2.13): no agent can be forced to buy or sell more than he wants to (though he may be forced to buy or sell less), neither can he be forced to change the sign of his intended trade (i.e. to buy if he wants to sell or conversely):

19 $\qquad\qquad\qquad |y_{ih}| \leq |\tilde{y}_{ih}| \qquad \forall i,h$

20 $\qquad\qquad\qquad y_{ih} \cdot \tilde{y}_{ih} \geq 0 \qquad \forall i,h.$

(Voluntary exchange has been assumed in this book as in almost all TER literature.)

(iii) *Orderly market* (Cf. **2.14**) If on a market there is positive (negative) aggregate effective demand, then for each agent his unfulfilled effective demand is of the same sign:

21
$$\sum_i \tilde{y}_{ih} \geq 0 \Rightarrow y_{ih} \leq \tilde{y}_{ih}$$
$$\sum_i \tilde{y}_{ih} \leq 0 \Rightarrow y_{ih} \geq \tilde{y}_{ih}$$
$\forall i$

This implies that on a balanced market each agent realizes his effective demand:

$$\sum_i \tilde{y}_{ih} = 0 \Rightarrow y_{ih} = \tilde{y}_{ih} \quad \forall i$$

as does any agent on the short side of the market. We have seen models both for orderly and for orderless markets. With *Benassy* the existence of equilibrium is also proved without assuming **21** to hold.

The above properties of the rationing scheme, whether (iii) is included among them or not, admit a large variety of applicable schemes:

Queueing: All the buyers (or all the sellers) are ordered in a queue. Agents at the head of the queue realize their effective demand; those at the tail realize 0 trade.

Uniform rationing: All the buyers (or all the sellers) perceive the same bound on a market, and realize their effective demand if it is smaller (in absolute value) than this bound, and realize the bound in the opposite case.

Proportional rationing: All the buyers (or all the sellers) can realize the same fraction (e.g. the half) of their effective demand. This is an example of a *manipulable* rationing scheme, any agent can improve his position by demanding more. If the agents perceive this scheme, the system may become unstable.

All these rationing schemes might have been formalized, the formalization depending on whether we assume a market to be orderly or orderless.

22 Perceived constraints revisited

The constraint vectors s_i and b_i for agent i are determined by the rationing scheme, from which the agent perceives a signal. (The question of how this signal transmission is carried out will be discussed soon.)

In the case of an *orderly market*, the following conditions define the perceived constraint:

a) If market h is in equilibrium, nobody perceives a constraint on this market

23 $$\sum_i \tilde{y}_{ih} = 0 \Rightarrow s_{ih} = -\infty, \quad b_{ih} = +\infty \qquad \forall i.$$

b) If market h is a seller's market, then sellers are not constrained, but buyers perceive an upper bound determined by the rationing scheme

24 $$\sum_i \tilde{y}_{ih} \geq 0 \Rightarrow \begin{cases} s_{ih} = -\infty, & \forall i \\ b_{ih} = y_{ih} & \text{if } \tilde{y}_{ih} > 0 \\ b_{ih} = +\infty & \text{if } \tilde{y}_{ih} \leq 0. \end{cases}$$

c) If market h is a buyer's market, then buyers are not constrained and the sellers receive a (negative) lower bound:

25 $$\sum_i \tilde{y}_{ih} < 0 \Rightarrow \begin{cases} b_{ih} = +\infty, & \forall i \\ s_{ih} = y_{ih} & \text{if } \tilde{y}_{ih} < 0 \\ s_{ih} = -\infty & \text{if } \tilde{y}_{ih} \geq 0. \end{cases}$$

In the case of an *orderless* market the generation of the perceived signal becomes more complicated, and both buyers and sellers may receive finite and binding signals on the same market. Discussion of this case is omitted here, but the interested reader may consult Benassy's book.

26 The K-equilibrium (BENASSY, 1982)

The K-equilibrium associated with a fixed price vector p and rationing scheme f_{ih} (h = 1,2,...,n; i = 1,2,...,m) is a set of effective demand vectors $\tilde{y}_i$, transaction vectors y_i and quantity constraints s_i, b_i, such that for all i:

(i) $\tilde{y}_i$ is determined as described in section **9**, given $\omega_i(\cdot)$, $\bar{v}_i$, $\bar{\mu}_i$ and **p** as data, s_i and b_i as below.

(ii) $y_i = f_i(\tilde{y}_1,...,\tilde{y}_n)$ (See **16**.)

(iii) s_i and b_i are determined as described in section **22** from y_i.

> A K-equilibrium may be viewed intuitively as a fixed point of the following *tâtonnement process in quantities*: Assume that all agents have expressed effective demand on all markets....From these we obtain perceived constraints, [and]...on the basis of these perceived constraints, the agents will determine a new set of effective demands...and so on. A K-equilibrium is reached when these new effective demands are the same as the former. (p. 77, my italics).

If **p** is positive, the ω_i's are continuous and strictly concave, and the rationing scheme is continuous and non-manipulable, then a K-equilibrium exists. Benassy proved this theorem using Brouwer's fixed point theorem (p. 81). In Figure 10 I depict the above tâtonnement process in a block diagram for sake of comparison with previous diagrams.

27 *An attempt at interpretation: the story of the quantity auctioneer*

So far I have reported on Benassy's model, as the most elaborated and frequently quoted representative of TER, and I hope I have done this without essential bias. Henceforth, I will try to interpret in my own words the processes implied by the model. Since, as quoted above, we have to deal with a tâtonnement process, I will personify it by a "quantity auctioneer" called Q.A., although advocates of TER might not approve of this personification. Let us nonetheless listen to Q.A.'s story.

On the morning of the transaction day someone enunciates the prices which will remain valid all day. Thereupon each agent expresses his nominal demand vector (supply = negative demand), and communicates this vector to Q.A. He then compares aggregate supply with demand in each market and announces his *first trial rationing* and transmits signals to the buyers and sellers for whom the rationing scheme has produced a *binding* constraint, in the form of a bound. Thereafter they again maximize utility and form their *first effective*

Figure 11

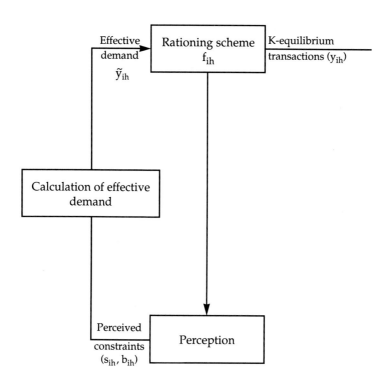

Block diagram of rationing

demand to be sent again to Q.A. Then a repetition of this information process goes on and on, and stops only when two consecutive sets of effective demand vectors are the same (i.e. the K-equilibrium is attained). Then (and only then) Q.A. announces that the tâtonnement phase has come to an end, transactions can be carried out, production and consumption can start and these continue until tomorrow when the story will be restarted with new prices.

28 Comments on the Q.A. story

As you might have noticed, the story is in a close parallel with the price auctioneer's story save for one difference. Quantity signals flow here in both directions, effective demands from the agents to Q.A. in one direction and bounds in the opposite direction. While the Walrasian auctioneer's output signal was the same price vector for everybody, Q.A. sends personalized messages to each agent.

As I reported under 3, GET has recently developed up to the point that the auctioneer could have been eliminated, production and consumption included. As far as I know, TER has not yet reached this stage; the problem seems to me considerable but not insoluble.

The first step in this direction would be to show that the quantity tâtonnement process is convergent and the resulting K-equilibrium state is stable. Stability results formed an integral part of GET early on (SAMUELSON, 1941); in Benassy's book the word "stability" does not even occur. (See, however, LAROQUE, 1981 for local results.) MALIN-VAUD (1977) is more explicit:

> To ask for a satisfactory answer to all these questions within a dynamic context is first to require a very extensive knowledge and, second, to suggest the elaboration of a model that will be very complex to handle....To rely on a general equilibrium formalization is to accept a short-cut, i.e. the consideration of only those equilibrium states that would result from dynamic adjustment. In other words, one assumes that individual decisions have had the time to adjust to each other so as to be mutually consistent. (pp. 6-7).

I do not challenge this argument as being justifiable in the context where TER is applied by Benassy, Malinvaud and others in macroeconomics. It is to be seen, however, that non-Walrasian economics (including TER as well as our approach) still has a long way to go to catch up with GET.

In the closing part of this Chapter I will try to compare the results of TER with my control theory approach and assess the differences from different points of view.

29 *First comparison: purposes and motivations*

The self-declared purpose of TER is, as I mentioned, to develop a microeconomic theory which lays the foundation of Keynesian macroeconomics, which in particular explains the most conspicuous lasting disequilibrium phenomenon in capitalistic economies —unemployment—and reveals the factors that effect its change.

It is not part of my intention (nor of anybody of the school I claim to belong to) to support Keynesian Theory, nor to challenge it. The socialist economy in which we live struggles with a series of "disequilibria" mostly in the form of shortages, including a labour shortage (KORNAI, 1980). *One* of the reasons why socialist economies get into functional disorders was identified long ago as attributable to insufficiently adaptive mechanisms. On the epistemological level this amounts to the lack of an explanatory theory of economic mechanism. This is what we are trying to contribute to.

The final purposes of the two research trends are thus considerably different, but both have a common interim purpose, i.e. to explain non-Walrasian phenomena.

30 *Second comparison: the contents of the quantity signals*

TER and our approach agree with each other in that, in contrast with GET, quantity signals play the predominant role in the short run adjustment process to a non-Walrasian steady state.

However, there is a marked difference in the contents of the signals. In TER the important quantity signals are: the perceived constraints and the effective demands. While perceived constraints can be given (after trade) a sufficiently verisimilar interpretation, their interpretation before trade goes via the quantity rationing story which is less realistic. And if we come to the notion of effective demand, its generation process is an ingenious construction requiring a series of operations whose original can hardly be found in real-life economy.

My primary objection to Q.A. is not that he resides on the same Olympus as the Walrasian auctioneer, but that his omniscience consists of an artifact: the effective demand.

Our quantity signals (on output stocks, input stocks, orders) are generated in a natural way; they are directly observable and taken from the everyday business practice. If we were considering only this, our approach would seem superior. But let us not forget another set of signals: the norms. Even if they are less mysterious, I have to admit that so long as they appear as celestial signs, without a full theory as to whence and how they are formed and changed, I must be more modest in the superiority claim.

It is interesting to note that even those who think in the terms of TER sometimes touch upon the possibility of our approach. GRANDMONT (1977) in his critical remarks writes:

> Indeed, in many cases, a buyer who is prevented from purchasing a commodity in the current period, will place an *order* to get the same commodity at some unspecified later date....In such a case, sellers have some information about the size of the excess demand for some commodity by looking at the size of purchases on the corresponding futures market....Other relevant indicators are the *level of stocks* in the case of durable goods, or leftovers otherwise....In the light of all this information, the traders would forecast supplies and demands for the future periods....This seems to be a realistic way of *modelling the flows of information which actually take place in our economies*. A difficult but rewarding project would be to study a model along these lines and to look at its dynamic implications. (p. 564, my italics)

The same idea appears also with MALINVAUD (1977):

> The immediate impact of changes in demand or supply is to be found in *order-books*, waiting lines, *inventories*, delivery dates....Such quantitative adjustments are the first signals of changes in the demand—supply relationship. (p. 9, my italics)

These quotations urge an approach which would be very similar to our own, but which — as far as I know — has not yet been pursued within TER.

31 *Third comparison: decision and information structure, the organizational setup*

Within GET, whether in its classical form or in novel variants, the price formation mechanism is a coordinated one, but once these (interim) signals have been formed and perceived by the agents the rest of the decisions—those on real activities—are uncoordinated. This setup drastically changes with rationing. Whenever a perceived constraint is binding (and whether for an agent it is binding or not), the amount which can be bought or sold is not decided by the agents who participate in the transaction, but must be coordinated by the rationing agent who knows all the effective demands.

We have placed a greater emphasis on uncoordinated processes (in Part Two), and whenever a coordinator appears (in Part Three), it can be identified as an abstract replica of some institution (Price Office, Production Office) which actually exists in certain economic systems.

Behind this organizational difference we can discover a deeper theoretical controversy. GET and TER both aim at explaining *market processes*; the area of their interest begins and ends on the market. In my approach it is the *non-market* processes which play the dominant role, specific market processes appear only accidentally. I am suggesting neither that the role of the market in economic control can be assumed away, nor that the study of its functioning has already been developed to such an extent that only gnawed bones remain. However, I am claiming that there is something outside the market which is also important in economic control, and which has been largely neglected by theoretical economists. Looking at the different approaches from this aspect they appear as *supplementary rather than competitive*.

32 *Fourth comparison: the extension and generality of the apprehended real processes*

As a consequence of the above difference in the approach to the control mechanism, the real processes apprehended are not co-extensive in the two approaches either.

In TER the agents (firms and households) appear in the market as owners of commodities and their real activity consists in exchanging them (in most but not all models with the mediation of money). Production and consumption are pushed into the background but the labour market fits well into the frame. The exchange process itself is treated on a rather high level of generality. There is no restriction with respect to the number of commodities and agents, the same commodity may be demanded and supplied by many of them. Whenever TER is applied to a productive firm or a household, the corresponding production possibility sets, production functions and utility functions are arbitrary (subject to continuity and convexity assumptions) and can be specified in different (including nonlinear) forms.

In my approach, on the other hand, in the representation of the real processes the production aspect is at the forefront and the transfer of commodities a close second. Consumption (a part of end use) is relegated to the "environment" affecting the real processes but not itself explained within the model. Labour as a non-storable commodity cannot be appropriately modelled. Generality is severely restricted by the specifications of the models. Each commodity is produced by only one producer and a unique technology, material inputs are proportional to output, productive capacities are fixed.

Thus from the comparison of the two approaches we conclude that, as regards *extension*, the difference is in the *shift of emphasis from exchange to production*. With respect to *generality* of formalization TER, inheriting the mathematical refinement of GET, is definitely superior.

33 Summary of the comparisons

The two approaches to modelling non-Walrasian economics were compared with respect to a set of arbitrarily selected criteria. They turned out to be related to each other at several points and complementary at others. They deviate in important economic principles and use different mathematical formalisms. TER stands out as using a higher and more general formalization of assumptions and conclusions at the expense of artificial constructions. We have tried to replace them with more natural ones. And we keep trying to develop these ideas, to widen their scope and generality.

BIBLIOGRAPHICAL NOTES TO PART THREE

The introduction to the Laplace transformation technique in *Chapter 15* is mostly standard textbook material. For a concise treatment see, e.g., Appendix B in ZADEH - DESOER (1963) or CSÁKI (1977), for a more detailed one see DOETSCH (1961) or SMITH (1966).

The equivalence concept for controllers as studied in *Chapter 16* has a parallel in the systems equivalence concept (ZADEH - DESOER, 1963). My conceptualization is closest to that of MCFARLANE (1971), but deviates from it by distinguishing the role of the external effects from that of the command signals. Although I have treated the equivalence notion in the context of a linear system (in terms of $\mathcal{L}$-transforms) its extension to nonlinear control systems is straightforward.

Chapter 17 is a revised version of what I wrote in Chapter 9 of NPC. The presentation has been simplified and made more transparent by the omission of an ambiguous labelling process, whose formalization was illusory.

The part of *Chapter 18* dealing with the rethinking of my approach in a historical context was inspired by the critical remarks of M. Augusztinovics.

The bibliographical associations of *Chapter 19* are included in the main text. Whenever I refer in the text to "our approach" it is to be understood that it subsumes not only my present work and its predecessors but also that of my fellow researchers. Among them the contributors to NPC, who were listed in the Bibliographical notes to Part Two, should be mentioned first. Other relevant contributors are LIGETI - SIVÁK (1978), KORNAI - WEIBULL (1978), SIMONOVITS (1981), KORNAI (1984).

In my treatise I have not dealt with the macroeconomic implications of this circle of ideas. KORNAI (1980) offers the most outstanding contribution. Further references are KORNAI (1982), SIMONOVITS (1982), KORNAI - SIMONOVITS (1986). A comparative analysis with the macroeconomic implications of TER has been provided by HARE (1982).

REFERENCES

AE: see KORNAI (1971).
ALLAIS, M. (1981): *La théorie générale des surplus.* Paris, Institut de Sciences Mathématiques et Economiques Appliquées. Cahiers "Economies et Societés."
ARROW, K.J. - G. DEBREU (1954): "Existence of an Equilibrium for a Competitive Economy." *Econometrica* **22** (265-290).
ARROW, K.J. - F.H. HAHN (1971): *General Competitive Analysis.* San Francisco, Holden Day.
ARROW, K.J. - L. HURWICZ (1960): "Decentralization and Comptutation in Resource Allocation." In PFOUTS (1960).
ARROW, K.J. - M.D. INTRILIGATOR Eds. (1982): *Handbook of Mathematical Economics.* Amsterdam, North-Holland.
AUBIN, J.P. (1981): "A Dynamical, Pure Exchange Economy with Feedback Pricing." *J. Econ. Behavior and Organization* **2** (95-127).
AUBIN, J.P. - A. CELLINA (1983): *Differential Inclusions.* Berlin, Springer.
BARRO, R.J. - H.I. GROSSMAN (1971): "A General Disequilibrium Model of Income and Employment." *American Econ. Review* **61** (82-93).
BARRO, R.J. - H.I. GROSSMAN (1976): *Money, Employment and Inflation.* Cambridge, Cambridge Univ. Press.
BELLMAN, R. (1953): *Stability Theory of Differential Equations.* New York, McGraw-Hill.
BELSLEY, D.A. (1969): *Industry Production Behaviour: The Order - Stock Distinction.* Amsterdam, North-Holland.
BENASSY, J.P. (1973): "Disequilibrium Theory." Unpublished Ph.D. Dissertation. (Hungarian translation in *Szigma*, 1974).
BENASSY, J.P. (1977): "On Quantity Signals and the Foundations of Effective Demand Theory." *Scandinavian J. of Economics* **79** (147-168).
BENASSY, J.P. (1982): *The Economics of Market Disequilibrium.* New York, Academic Press.
BODEWIG, E. (1959): *Matrix Calculus.* Amsterdam, North-Holland.
CHIKÁN, A. Ed. (1982): *The Economics and Management of Inventories.* Amsterdam, Elsevier.

CLOWER, R.W. (1965): "The Keynesian Counterrevolution: A Theoretical Appraisal." Chapter 5 in HAHN - BRECHLING (1965).

CSÁKI, F. (1977): *State-Space Methods in Control Systems.* Budapest, Akadémiai Kiadó.

CSIKÓS-NAGY, B. Ed. (1984): *The Economics of Relative Prices.* London, MacMillan.

DEBREU, G. (1959): *Theory of Value.* New York, Wiley.

DERAKHSAN, M. (1978): "Bibliography on Application of Systems and Control Theory in Economic Analysis." Mimeo. Oxford University Engineering Laboratory Report No. 1267/78.

DRAZEN, A. (1980): "Recent Developments in Macroeconomic Disequilibrium Theory." *Econometrica* **48** (283-306).

DRÈZE, J. (1975): "Existence of an Exchange Equilibrium under Price Rigidities." *International Economic Review* **16** (301-320).

DOETSCH, G. (1961): *Guide to the Applications of Laplace - transforms.* Princeton, N.J., D.U. Nostrand.

ELGERD, O.I. (1967): *Control Systems Theory.* New York, McGraw-Hill.

FISHER, M.F. (1983): *Disequilibrium Foundations of Equilibrium Economics.* Cambridge, Cambridge Univ. Press.

GALE, D. (1960): *The Theory of Linear Economic Models.* New York, McGraw-Hill.

GANTMACHER, F.R. (1959): *Theory of Matrices.* New York, Chelsea.

GEYER, W. - W. OPPPELT, Eds. (1957): *Volkswirtschaftliche Regelungsvorgänge im Vergleich zu Regelungsvorgängen in der Technik.* Műnchen.

GOODWIN, R.M. (1963): "Static and Dynamic Linear General Equilibrium Models." Reprinted in GOODWIN (1983).

GOODWIN, R.M. (1983): *Essays in Linear Economic Structures.* London, MacMillan.

GRANDMONT, J.M. (1977): "The Logic of the Fix-Price Method." *Scandinavian J. of Economics* **79** (170-186).

GRANDMONT, J.M. (1982): "Temporary General Equilibrium Theory." Chapter 19 in ARROW - INTRILIGATOR (1982).

HAHN, F.H. (1962): "A Stable Adjustment Process for a Competitive Economy." *Review of Economic Studies* **29** (62-65).

HAHN, F.H. (1978): "On Non-Walrasian Equilibria." *Review of Economic Studies* **45** (1-17).

HAHN, F.H. - F. BRECHLING, Eds. (1965): *The Theory of Interest Rates.* London, MacMillan.

HAHN, F. H. - T. NEGISHI (1962): "A Theorem on Non-tâtonnement Stability." *Econometrica* **30**.
HARE, P.G. (1982): "Economics of Shortage and Non-Price Control." *J. of Comparative Economics* **6** (406-425).
HICKS, J. (1965): *Capital and Growth*. Oxford, Clarendon Press.
HICKS, J. (1974): *The Crisis in Keynesian Fconomics*. Oxford, Basil Blackwell.
JANSSEN, J.M.L. - L.F. PAU - A. STRASZAK, Eds. (1981): *Dynamic Modelinq and Control of National Economies*. Oxford, Pergamon Press.
KAPITÁNY, Zs. (1982): "Dynamic Stochastic Systems Controlled by Stock and Order Signals." In CHIKÁN (1982).
KAWASAKI, S. - J. MCMILLAN - K. F. ZIMMERMANN (1982): "Disequilibrium Dynamics: An Empirical Study." *American Economic Rev.* **72** (992-1003).
KORNAI, J. (1971): *Anti-Equilibrium*. Amsterdam, North-Holland.
KORNAI, J. (1980): *Economics of Shortage*. Amsterdam, North-Holland.
KORNAI, J. (1982): *Growth, Shortage and Efficiency*. Oxford, Basil Blackwell.
KORNAI, J. (1984): "Adjustment to Price and Quantity Signals in a Socialist Economy." In CSIKÓS - NAGY (1984).
KORNAI, J. - B. MARTOS (1973): "Autonomous Control of the Economic System." *Econometrica* **41** (509-528).
KORNAI, J. - B. MARTOS, Eds. (1981): *Non-Price Control*. Amsterdam, North-Holland.
KORNAI, J. - A. SIMONOVITS (1977): "Decentralized Control Problems in Neumann Economies." *J. Economic Theory* **14** (44-67).
KORNAI, J. - A. SIMONOVITS (1986): "Investment, Efficiency and Shortage: A Macro Growth Model." *Matekon* **22** (3-29).
KORNAI, J. - J.W. WEIBULL (1978): "The Normal State of the Market in a Shortage Economy: A Queue Model." *Scandinavian J. of Economics* **80** (375-398).
LACKÓ, M. 1982 : "Magatartási szabályok a beruházások ágazatközi elosztásában." (Behavioural Rules in the Intersectoral Allocation of Investments.) *Közgazdasági Szemle* **28** (848-862).
LANCASTER, P. (1969): *Theory of Matrices*. New York, Academic Press.
LANGE, O. (1965): *Wstęp do cybernetyki ekonomicznej*. (Introduction to Economic Cybernetics). Warszawa, Państwowe Wydawnictwo Naukowe.

LAROQUE, G. (198]): "On The Local Uniqueness of the Fixed Price Equilibria." *Review of Economic Studies* **48** (113-129).
LEIJONHUFVUD, A. (1968): *On Keynesian Economics and the Economics of Keynes.* Oxford, Oxford Univ. Press.
LIGETI, I. - J. SIVÁK (1978): *Növekedés, szabályozás és stabilitás a gazdasági folyamatokban.* (Growth, Control and Stability in Economic Processes.) Budapest, Közgazdasági és Jogi Könyvkiadó.
MCFARLANE, A.G.J. (1971): "Linear Multivariable Feedback Theory: A Survey." In SCHWARCZ (1971).
MALINVAUD, E. (1977): *The Theory of Unemployment Reconsidered.* Oxford, Basil Blackwell.
MARTOS, B. (1981): "Non-Price Control (Report on a Research Trend)." In JANSSEN - PAU - STRASZAK (1981).
MARTOS, B. (forthcoming): "Viable Control Trajectories in Linear Systems."
MELSA, J.L. - D.G. SCHULTZ (1969): *Linear Control Systems.* New York, McGraw-Hill.
MESAROVIC, M. - D. MACKO - Y. TAKAHARA (1970): *Theory of Hierarchical Multilevel Systems.* New York, Academic Press.
MORGENSTERN, O. Ed. (1954): *Economic Activity Analysis.* New York, Wiley.
NEGISHI, T. (1962): "The Stability of a Competitive Economy." *Econometrica* **30** (635-669).
NPC: see KORNAI - MARTOS (1981).
OSTROWSKI, A.M. (1974): "Subdominant Roots of Non-negative Matrices." *Linear Algebra* **8** (179-184).
PFOUTS, R.W. Ed. (1960): *Essays in Economics and Econometrics.* Chapel Hill, Univ. of N. Carolina Press.
PHILLIPS, A.W. (1954): "Stabilization Policy in a Closed Economy." *Economic Journal* **64** (290-323).
ROSEN, R. (1985): *Anticipatory Systems.* Oxford, Pergamon Press.
ROSENBROCK, H.H. (1970): *State Space and Multivariable Theory.* New York, Wiley.
SAMUELSON, P.A. (1941): "The Stability of Equilibrium: Comparative Statics and Dynamics." *Econometrica* **9** (97-120).
SAMUELSON, P.A. (1947): *Foundations of Economic Analysis.* Cambridge, Harvard Univ. Press.

SCHWARZ, H. Ed. (1971): *Multivariable Technical Control Systems. Proceedings of the 2nd IFAC Symposium, Düsseldorf.* Amsterdam, North-Holland.

SIMON, H.A. (1952): "On the Application of Servomechanism Theory in the Study of Production Control." *Econometrica* **20** (247-268).

SIMONOVITS, A. (1981): "Maximal Convergence Speed of Decentralized Control." *J. Economic Dynamics and Control* **3** (51-64).

SIMONOVITS, A. (1982): "Buffer Stocks and Naive Expectations in a Non-Walrasian Dynamic Macromodel: Stability, Cyclicity and Chaos." *Scandinavian J. Economics* **84** (571-581).

SMITH, M.G. (1966): *Laplace Transforms Theory.* Princeton.

TSUKUI, J. (1968): "Application of a Turnpike Theorem to Planning for Efficient Accumulation: an Example for Japan." *Econometrica* **36**.

TUSTIN, A. (1953): *The Mechanism of Economic Systems.* London, Heinemann.

UZAWA, H. (1962): "On the Stability of Edgeworth's Barter Process." *International Economic Rev.* **3** (218-232).

WALRAS, L. (1874-77): Eléments d'économie politique pure. Lausanne, Corbar.

ZADEH, L.A. - C.A. DESOER (1963): *Linear System Theory.* New York, McGraw-Hill.

INDEX

Actuating signal 46, **60**, 89
Admissible input set **71**, 126, 108
Arrow-Debreu Theory **216**
Auctioneer 18, 31, **216**, 224

Backlog of orders 24, 27, **114**, 130, 139
Basic set **182**, 189
Behavioural equations 85, **98**
Borderline case **68**-69
Bounded input **68**, 69, 77
Bounded time-function **65**
Budget constraint **219**

Center **73**, 203
Characteristic equation **93**, 104, 154
Command signal **44**, 46, 51
Command transfer **179**
Commercial sector **137**, 138
Commercial stock **138**
Commodity **12**, 19, 20, 27, 37, 75, 138
Compatible input **74**
Compatible state **74**
Constant-input equilibrium state **58**
Constant-input steady state **59**, 106, 125, 134, 143, 155
Control agent **12**
Control by norm **23**-26, 65, 79
Control circuit 32, **46**, 53, 60-62, 162, 177, 178
Control organizations **12**-14, 35, 39-41
Control processes **12**-15, 18, 32, 34, 36-39, 46, 48, 49, 129
Control space **55**, 71
Control sphere 13, 14, **17**, 48, 49
Control system 15, 16, 43, **44**, 50, 51, 62, 63, 68, 69, 79, 178
Control theory 8, **11**, 24, 43, 44, 49, 53

Control unit **12**, 34, 37
Control vector **55**
Controlled subsystem **43**, 49, 50, 60, 61
Controlled variable **44**, 46
Controller 32, 36-38, **43**, 44, 46, 48, 50, 51, 60, 61, 63, 69, 130, 163, 167, 193
Coordinated **34**, 39, 41
Cube norm **66**

Damping exponent **90**, 163
Decentralized 31, **34**, 40
Degree of stability **71**
Diagonal matrix 66, **192**
Discrimination **20**, 21, 79
Distinct eigenvalues **86**
Disturbance transfer **179**
Duality **203**
Dynamics **16**, 17, 25, 41

Effective demand **218**
Eigenvalue **66**-68
Eigenvector **66**
End use 12, 15, 16, **84**, 88, 116
Equilibrium state 17, **58**, 59, 69, 80
Equivalence **181**, 189, 195
Equivalence classes **181**
Equivalence conditions **185**
Equivalent controllers **183**, 191
External effect **12**, 44, 46, 54, 61, 178

Feedback 32, **46**, 60
Fiduciary goods **13**, 14

General Equilibrium Theory (GET) **215**

History 25, **211**

Identity transfer element **37**
Initial state 16, **57**, 58, 60, 67, 127

Input matrix 55
Input signal 37, 50, **53**, 57, 60
Input stock 85
Input trajectory 57, **69**, 71, 75
Instability **67**, 69
Introactive 37
Introspective 37, 38
Inventory 32, 36, 70, **97**
Irreducibility 86
Irreducible matrix 192
Irrelevant variables 101

$\mathcal{L}$-transform (see Laplace transform)
Laplace-transform 53, **174**, 182
Laplace-transformation **173**, 186, 188, 194
Leontief economy **15**, 16
Linear system **49**, 58, 59, 69

Manipulating variable 37, **44**, 46, 48
Manipulator **36**, 37, 50, 60
Market **18**-21, 23, 27, 39, 41, 79, 130, 162, 222, 223
Market frictions 20
Matrix norm **64**, 65
Modal condition number **71**, 72
Modal matrix **66**, 72
Monetary processes **13**, 14
Monocentric control 40

Natural frequency 90
Non-diagonal matrix 192
Norm **23**-26, 64, 65, 72, 73, 77, 79, 80, 162

Open ball 73
Open-loop control **44**, 46, 48
Optimization **23**, 25, 28
Order placements **114**-115, 140, 162
Order signal **113**, 135
Orderless market **20**, 130
Orderly market **19**, 20, 79, 222
Output matrix 55
Output signal 36, 37, 50, **53**
Output space 55
Output stock 24, 32, **85**, 150, 209
Output vector **55**, 56

Partial process 32

Partially centralized control 39
Perceived constraint 218
PID element 54
Polycentric control 40
Price office 41, **199**, 204
Price signal 39, **147**
Pricing matrix 193
Production 6, 12, 16, 24, 27, 28, 30, 36, 44, 70, **84**, 100, 130, 140, 197
Production decision matrix 193
Production office 201
Productivity 86

Quantity auctioneer (Q.A.) **224**

Radius **73**, 123
Rationing scheme 18, **218**, 221, 223
Real agent 12
Real organizations **12**-14, 34, 35, 38-40, 207
Real processes **12**, 14, 15, 18, 34, 36, 39, 48, 114, 129, 138, 139, 149, 150, 167, 182, 186, 229
Real sphere **13**-17, 48, 63
Real unit **12**, 37
Reduced form **91**, 103, 119, 132
Reducible matrix 192
Retail market 142
Return transfer **85**, 107, 162
$r\mathcal{L}$v-positive **101**, 108-110, 127

Sensor **36**, 37, 50, 54, 60
Short-side rule **19**-21, 79
Signal **12**, 14, 18, 32, 34-39, 44, 46, 50, 51, 53, 57, 60, 97, 113, 135, 137, 147, 162
Signal generation 12, **32**, 34, 35, 38, 50
Signal transmission **34**, 35
Simple matrix **66**, 72
Simplified $\mathcal{L}$-transform **178**
Spectral decomposition 66
Spectral matrix 66
Spectrum 5, **92**, 119, 125, 134, 143
Spillover effect **219**
Stability 8, 17, 59, 63, **67**-69, 71, 75, 135, 162, 196
State equation **53**, 55-57, 60, 61, 66, 67

State space **55**, 71, 74
State trajectory **69**, 70, 75
State transition matrix **57**
State vector **55**, 56, 60, 62, 67, 68
Steady state **59**, 67, 68, 80, 106, 125, 134, 143, 155
Stock signal **97**, 135, 137
Stone age **211**
Sub-process **32**
Summation vector **88**
Supply-side price **148**
System matrix **55**, 67, 71, 72, 75

Tâtonnement 18, **216**, 217
Technology 12, 15, **85**
Theory of Equilibria with Rationing **215**
Time-invariant system **49**, 70
Time-variant system **49**
Trader **137**, 162
Transfer 12, 32, 35-38, 44, 46, 53-55, 60-62, **84**, 107, 117, 139, 151, 162,
Transfer element **32**, 35-37, 53-55, 60-62, 176, 177
Transfer matrix **177**

Value added **148**, 162
Vector norm **64**
Vegetative functioning 31, **35**, 80
Viability 8, 23, 49, 50, 63, 68, **69**, 71-75, 77, 80, 196
Viability set **71**, 73, 75, 80, 162
Voluntary exchange **19**, 221

Wholesale market **140**

Zero-input equilibrium state **58**, 80
Zero-input steady state **59**, 80